Environmental Leadership Capacity Building in Higher Education

Takashi Mino • Keisuke Hanaki
Editors

Environmental Leadership Capacity Building in Higher Education

Experience and Lessons from Asian Program for Incubation of Environmental Leaders

 Springer

Editors
Takashi Mino
Graduate School of Frontier Sciences
The University of Tokyo
Kashiwa, Chiba, Japan

Keisuke Hanaki
Department of Urban Engineering
Graduate School of Engineering
The University of Tokyo
Bunkyo-ku, Tokyo, Japan

ISBN 978-4-431-54339-8   ISBN 978-4-431-54340-4 (eBook)
DOI 10.1007/978-4-431-54340-4
Springer Tokyo Heidelberg New York Dordrecht London

Library of Congress Control Number: 2013935477

Printed on acid-free paper

Springer is part of Springer Science+Business Media (www.springer.com)

Preface

Sustainability is an important keyword in the design of future society, and the environmental dimension is one of the most critical aspects of sustainability. Environmental problems typically involve many stakeholders and have a complex nature with many uncertainties. This makes it difficult to come up with a consensus among the stakeholders in tackling these problems. In this decision-making process, we need leaders. They should be able to see problems holistically, to understand the sociocultural and human factors of the concerned community, and to propose a new framework that may mitigate the existing problems and provide an alternative path for a sustainable new paradigm. It is an essential duty for the whole society, especially for higher education institutions, to foster such leaders.

In recent decades, The University of Tokyo (UT) has been keen on the issue of the environment and/or sustainability and has endeavored to establish new multidisciplinary schemes that focus on these issues. An outcome of such efforts was the launch of the Division of Environmental Studies in 1999, which aimed to solve environmental problems through interdisciplinary collaboration among relevant disciplines. In the same year, the Swiss Federal Institute of Technology (ETH), the Massachusetts Institute of Technology, and UT came together in the Alliance for Global Sustainability (AGS) to discuss the possibility of organizing a summer program on sustainability, which materialized the following summer as the Youth Environmental Summit (later renamed Youth Encounter on Sustainability, YES). UT brought this movement over to Asia and initiated another summer program called the Intensive Program on Sustainability (IPoS) in 2004. YES and IPoS were experimental projects to establish pedagogy and materials for sustainability education at a higher education level by making use of the diversity of students as well as that of instructors.

As a consequence of these educational challenges with a transdisciplinary nature, a formal graduate program, the Graduate Program in Sustainability Science (GPSS), was established under the Division of Environmental Studies at UT in 2007. The Integrated Research System for Sustainability Science (IR3S), an institution developed in UT in 2005 to coordinate and develop research collaboration on sustainability among major Japanese universities, played a key role in the establishment and development of GPSS and offered it strong support both financially and academically. GPSS, initially started as a Master's program, became a Ph.D. program in 2009. In 2011, GPSS was awarded an exclusive educational project from the Ministry of Education, Culture, Sports and Technology (MEXT), Japan, under the scheme of "Program for Leading Graduate Schools" and is now in the process of strengthening its Ph.D. components.

The Department of Urban Engineering (UE) was established in 1962 in the Faculty of Engineering at UT with the aim of dealing with complex urban issues in holistic ways and providing a strong practical basis for urban planning and environmental technology/management. Over the past 50 years, UE has gained an excellent reputation internationally and academically in relevant fields. One of the characteristics of their curriculum is a strong emphasis on studio and/or laboratory work. This enabled UE to develop diverse sets of teaching modules in case studies and experiential learning. Another point is that UE has been accepting students from all over the world since its early stages and has accumulated know-how in education for international students. In 2012, UE was accepted for MEXT's "Re-Inventing Japan Project" and is currently working to develop a new international collaboration scheme for student exchange.

The encounter of GPSS and UE brought about a unique educational challenge: The Asian Program for Incubation of Environmental Leaders (APIEL).

This book, consisting of eight chapters, is a summary of APIEL's four years of educational challenges. The structure of the book is as follows. The overall picture of APIEL is in Chap. 1, which describes how APIEL was established with its aims and core concepts. The objective of APIEL is to incubate an environmental leader who can resolve complex problems.

The concepts of the environmental leader, reviewed through a discourse on leadership, are defined in Chap. 2. The history of the development of environmental education and leaders is also discussed in this chapter. Some personal experiences from professional environmentalists are included as well.

In APIEL, future environmental leaders were incubated through compulsory courses which consisted of lectures and field exercise. "Environmental Challenges and Leadership in Asia" is one of the compulsory courses that was specially developed for APIEL. Detailed information about this course is given in Chap. 3.

Chapters 4–7 provide examples of APIEL's field exercises, namely, the Thailand Unit, the Oasis Unit, the GPRD Unit, and the Cambodia Unit, respectively. This series of chapters showcases a variety of field exercises and different perspectives. These field exercises were established by UT in collaboration with counterpart universities in other countries. The approach for establishing the field exercises, characteristics, lessons learned, and outcome of the field exercises is described in each chapter.

Finally, feedback from collaborating counterpart universities and alumni is reflected in Chap. 8.

As can be seen from the above history and the book itself, a huge amount of time, human resources, thought, and effort had been invested before APIEL started, and they are being invested continuously even now to develop, operate, and improve the educational scheme for environmental or sustainability education at the higher education level. We believe that this book will be a truly valuable milestone for those who are thinking of meeting the same kinds of educational challenges.

This book represents only a fraction of APIEL. More information and newsletters are available at http://www.envleader.u-tokyo.ac.jp/index_e.html.

Tokyo, Japan

Takashi Mino
Keisuke Hanaki

Acknowledgments

The Asian Program for Incubation of Environmental Leaders (APIEL) started in the year 2008 as a five-year project when it received "Strategic Funds for the Promotion of Science and Technology, Ministry of Education, Culture, Sports, Science and Technology, Japan." APIEL would never have been realized without this governmental financial support, and we would like to express our gratitude for being given this opportunity. APIEL thanks the two presidents of The University of Tokyo, Hiroshi Komiyama and Junichi Hamada, who served as leaders of this project.

Our thanks and appreciation also go to the Coca-Cola Educational & Environmental Foundation and to the Daiwa Securities Group Inc., who supported APIEL by providing generous financial aid. They enabled us to develop our activities on a larger scale, especially for strengthening the networks among the young professionals who participated in our education programs implemented inside and outside Japan.

APIEL enjoyed the great privilege of working with a huge number of people and organizations who supported us in various aspects of the establishment and implementation of our education program. Words are not adequate in offering our gratitude; nevertheless, APIEL would like to sincerely thank all those who were involved in our education programs, including those whose names are not mentioned in this acknowledgment.

APIEL would like to thank Springer Japan, Tokyo, for giving us this wonderful opportunity to publish our achievement in this book. We also express our appreciation to Ms. Izumi Ikeda, APIEL, for her great coordinating and editorial work.

Contents

Chapter 1
Asian Program for Incubation
of Environmental Leaders

Tomohiro Akiyama, Keisuke Hanaki, and Takashi Mino

Abstract This chapter outlines the features of the Asian Program for Incubation of Environmental Leaders (APIEL), including its objectives, core concepts and curriculum structure. APIEL is an educational program developed by The University of Tokyo that aims to foster environmental leaders, who have wide knowledge base, critical perspective, and a strong ethical stance. Those environmental leaders are expected to contribute to building environment-friendly and sustainable societies in the future in Asian countries. In addition, APIEL intends to create a collaborative network of higher education institutions in Asia with a view to tackling environmental issues by developing environmental leadership capacity.

Keywords Asia • Education • Environmental problem • Interdisciplinary • Interregional • Leadership • Stakeholder

T. Akiyama (✉)
Graduate Program in Sustainability Science, Graduate School of Frontier Sciences,
The University of Tokyo, Environmental Studies Building 334, 5-1-5 Kashiwanoha,
Kashiwa, Chiba 277-8563, Japan
e-mail: akiyama@k.u-tokyo.ac.jp

K. Hanaki
Department of Urban Engineering, Graduate School of Engineering, The University of Tokyo,
7-3-1 Hongo, Bunkyo-ku, Tokyo 113-8656, Japan
e-mail: hanaki@env.t.u-tokyo.ac.jp

T. Mino
Graduate School of Frontier Sciences, The University of Tokyo,
5-1-5 Kashiwanoha, Kashiwa, Chiba 277-8563, Japan
e-mail: mino@mw.k.u-tokyo.ac.jp

T. Mino and K. Hanaki (eds.), *Environmental Leadership Capacity Building in Higher Education*, DOI 10.1007/978-4-431-54340-4_1,

1.1 Introduction

The technological innovation and economic growth in the twentieth century have led to significant improvement of human welfare; however, they also caused unprecedented environmental problems. These problems include not only local/regional environmental pollution, but also global issues such as climate change, resource shortages, the energy crisis, and the loss of biodiversity. Environmental problem is complex because it relates not only to technical/physical issues but also social issues. In addition, even if an environmental problem appears peculiar to a region, it is closely connected to global issues. Thus, what is important to find solutions on current environmental problems is to integrate the scientific knowledge with findings from other perspectives. It is also important to develop human resources who hold a holistic view.

To foster environmental leadership in graduate students, the Asian Program for Incubation of Environmental Leaders (APIEL) was established in 2008 by The University of Tokyo. This educational program offers a curriculum that develops environmental leaders, who can make significant contributions to resolving global environmental problems as well as regional/local environmental problems in the twenty-first century.

APIEL is not a degree program but a certificate-awarding program originally developed through collaboration between the Graduate Program in Sustainability Science (GPSS) of the Graduate School of Frontier Sciences and the Department of Urban Engineering (UE) of the Graduate School of Engineering. This collaboration has given a unique strength to the curriculum because the former provides transdisciplinary approaches to APIEL, and the latter provides APIEL a solid academic base of environmental technology and management. In this way, APIEL is able to provide structured knowledge, practical skills, and experiences to students. APIEL particularly focuses on the environmental problems in Asia because Asia is the most populated region in the world and Asian countries at different stages of development are facing the challenge to build a sustainable society with limited natural resources.

1.2 Core Concepts

1.2.1 Characteristics of APIEL

To resolve the complex problems mentioned above and to contribute to sustainable development requires highly specialized leaders with wide knowledge, a critical perspective, and a strong ethical sense. APIEL's goal is to be an incubator for environmental leaders who have the necessary practical skills. In particular, APIEL puts emphasis on the following core concepts.

First, in the process of tackling environmental problems, it is important to recognize both universality and locality. Scientific bases and core environmental technologies can be universally applied in many cases but, in practice, specific local conditions must always be considered. Therefore, environmental leaders should be able to identify the problems within the *global* context based on universal knowledge, as well as understand and analyze the problems within the *local* context. They must be able to think clearly about cultural and social factors, local ecological and geographical characteristics, and interactions among the neighboring regions/communities, and to propose environment-friendly solutions and sustainable systems from a holistic point of view.

Second, respect for cultural and disciplinary diversity is a key concept of APIEL because these are essential for developing an efficient partnership among the various stakeholders. Collaboration among stakeholders is necessary to take on complex environmental problems, and this collaboration can only be achieved by understanding the views of different cultures and disciplines. In the APIEL curriculum, interactions among students, instructors, and stakeholders from varied cultural and disciplinary backgrounds are crucial—both in the classroom and in the field.

Third, hands-on experience and experiential learning should be the key pedagogy in the curriculum. APIEL introduced field exercises that provide students with opportunities to learn how environmental problems become complex and how different stakeholders are involved. Students learn how to deal with contradictions around problems through real experience with environmental cases and exposure to complex situations.

In addition, APIEL aims to create a resonant network of higher education institutions in Asia dealing with environmental research and education. Partners were identified during the development of the field exercises. As well, exchanges of information and the transfer of key experiences in teaching are essential for all similar institutions as there are a limited number of models available for environmental leadership education. The APIEL network should work as the platform for this type of collaboration.

1.2.2 Environmental Leadership

APIEL is striving to develop human resources who are able to play major roles as environmental leaders in various organizations in the world. APIEL expects environmental leaders to:

1. *Recognize* global and regional/local environmental problems and propose solutions to these problems using not only specialized professional knowledge and skills, but also inter-disciplinary thinking and approaches.
2. *Acquire* a balanced understanding of the knowledge, skills, and ways of thinking of the natural sciences as well as the humanities and social sciences.
3. *Refine* the ability in the field to make judgments, take action, and work in partnerships to resolve real-world environmental problems.
4. *Develop* the communication and leadership skills necessary to raise topics for discussion and to negotiate issues in several international as well as local situations.

These leaders may in the future play a key role in decision-making processes within different levels of society, including companies and NPOs, regional communities, specialist/professional groups, local and national governments and various international organizations. They will be expected to lead society in an environment-friendly and sustainable direction.

1.2.3 Resonance

Interactions among disciplines, stakeholders, and regions are essential for solutions of environmental issues. These interactions should bring benefits to all sides. These mutual positive influences can be called "resonance." And this "resonance" is one of the core concepts of APIEL.

Experiencing and learning from the various types of interaction are at the core of the program. Forming heterogeneous student groups for classes and conducting field studies with other universities helps to create resonance. Students from different backgrounds, such as engineering and the social sciences, are intentionally mixed together. Through this education program, the students learn to interact. In addition, faculty members or institutions form a resonance among themselves. An interaction resonance is also formed through alumni activities. The environmental leaders from several countries incubated by this program go back to their home countries, and then host the field study of younger students.

Forming an academic network among universities is an effective way to implement education for environmental leadership developments. Universities can exchange students and faculty members as well as information on academic programs. Such a network in Asia can form a strong base for collaborative education and research on Asian environmental issues. Interdisciplinary resonance, interregional resonance and alumni resonance are enhanced through these exchanges (Fig. 1.1). The detail information of each resonance will be explained in the later part.

1.2.3.1 Interdisciplinary Resonance

APIEL targets interdisciplinary resonance by providing an integrated, multidisciplinary curriculum as well as educational and research opportunities that promote mutual understanding among individuals in environmental fields. Scientific disciplines have been developed, widened, and deepened especially since the Industrial Revolution. Today, we need interaction among different disciplines. Environmental leaders are supposed to understand each discipline and make "bridges" across them to solve problems in the real world.

Figure 1.2 shows an example of disciplines involved in the occurrence and solution of environmental issues. Environmental deterioration appears as changes in water, air, soil, and the global environment. The natural sciences observe, analyze, and manage these changes. The impact of environmental change on health is a critical issue, and this is covered by the medical sciences. Many fields of engineering contribute to solving pollution problems or climate change through technologies such as wastewater treatment, energy-saving devices, or photovoltaic cells. Agriculture is influenced by, and, on the other hand, causes environmental

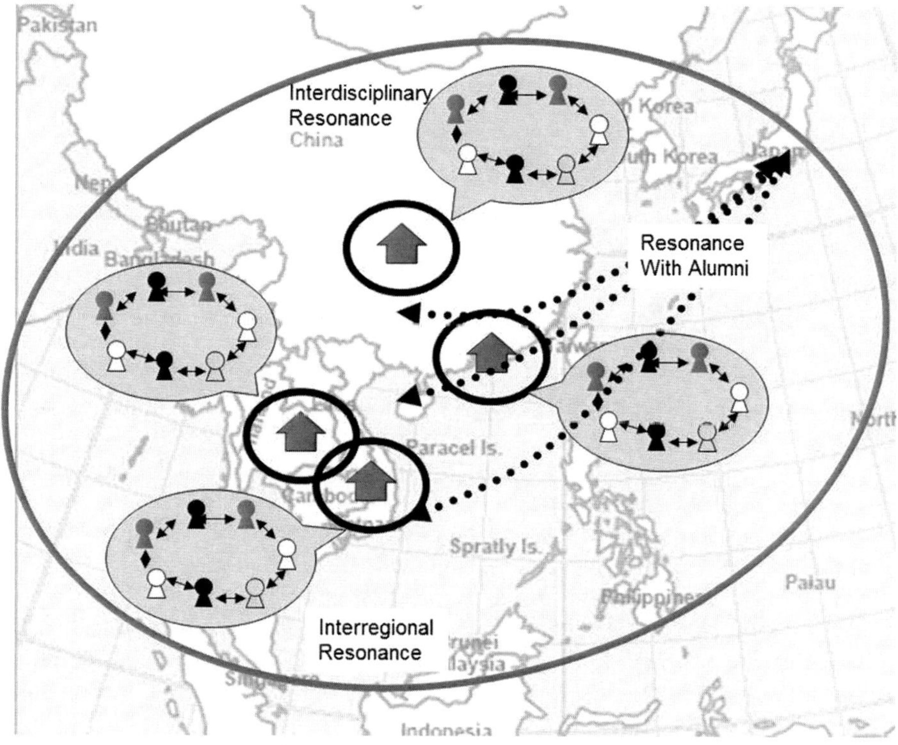

Fig. 1.1 Concept of resonance in educational program for environmental leaders

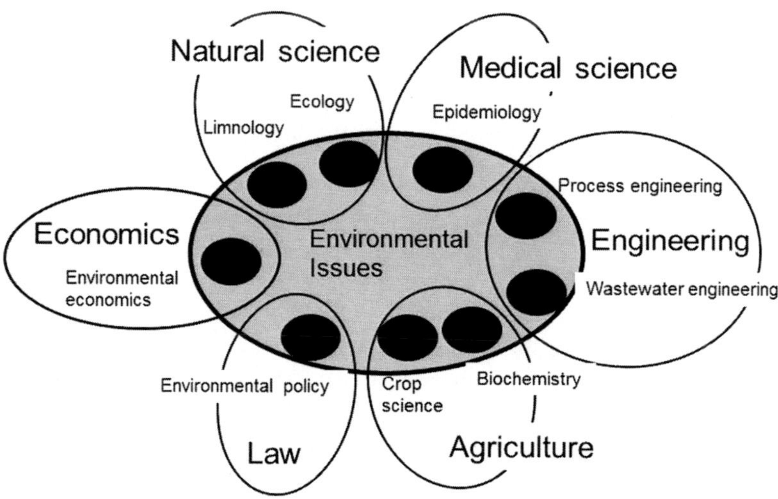

Fig. 1.2 Interdisciplinary characteristics of environmental issues

deterioration. Social sciences are also important to manage and control the environment. Environmental policy in the field of law and environmental economics in the field of economics are the examples.

Though these disciplines are necessary to identify and analyze environmental issues and propose solutions, their contribution is limited if each discipline is separated. Interaction among these disciplines is essential. For example, the wrong technology could be selected if an engineer does not consider local social and economic circumstances. On the other hand, the wrong technology can also be selected if the decision-making team has no expert in engineering and just considers the cost. There is a risk for official development assistance, where important funding negotiations take place between governments, when the recipient nation demands an expensive high-tech solution that requires experts to operate and maintain. The assisting nation then tries to cut the budget without considering operating and maintenance costs. The outcome is that expensive technology and machines are introduced but stop running within a short period. The role of technology experts who understand the social circumstances is essential in such a case. The natural sciences and engineering are superficially under a common discipline, but there are significant differences in basic concepts and methodology. The ultimate objective of the natural sciences is to find the truth; engineering is oriented to problem solving. Understanding every aspect of these different disciplines is impossible for one person. Instead, having the ability to interact with experts from different disciplines is essential for environmental leaders. In other words, environmental leaders must be coordinators among these different disciplines.

A holistic academic organization, such as a national academy, should play a fundamental role in promoting interdisciplinary academic activity. For example, the Science Council of Japan set up a committee on environmental science as a body that cuts across three major groups: social sciences and the humanities, life and medical sciences, and the natural sciences and engineering. However, real environmental issues cannot wait until disciplines merge before finding a solution for environmental issues. Therefore, for solving an individual issue without any delay, an environmental leader can coordinate the disciplines.

1.2.3.2 Interregional Resonance

Environmental issues have commonalty and diversity among countries or regions. The mechanisms of pollution or climate change are common among the regions. However, the history, the causes, and control of these problems are local issues. Effective technology for solutions is to some extent common, but the choice of technology depends on local circumstances.

Even for global environmental issues, such as climate change, reducing greenhouse gases depends on local conditions. The impact of climate change is a locally dependent issue. Analyzing environmental issues requires an understanding of environmental deterioration within the local context. Though data is gathered through surveys, much more valuable and relevant information is obtained through collaboration with local people, governments, or experts. The global environmental issues cannot be solved by only region-specific efforts. This APIEL targets interregional resonance by developing partnerships and forming networks with educational and research institutes throughout Asia.

1.2.3.3 Alumni Resonance

To solve real environmental problems, coordination and consensus among the diverse stakeholders are also essential. Stakeholders include citizens, and people from local governments, industry, and agriculture as well as others who are affected. The influence on future generations should be considered during decision making, though there is no representative for future generations. The coordination process for reaching a consensus requires many steps and much time. However, this process is essential for developing, implementing, and sustaining a solution. Otherwise, the solution cannot work properly and will need to be changed.

The APIEL also targets resonance with alumni by promoting interaction and information exchange and sharing among alumni. Alumni are expected to come back to the APIEL as instructors or to help refine the program's curriculum and design.

1.3 Curriculum Structure of APIEL

To foster students' environmental leadership, APIEL established a compulsory course called Environmental Challenges and Leadership in Asia, and a companion field exercise course. In the former, students learn, in an interactive way, about environmental leadership and various environmental problems in Asia. In the latter, practical issues are examined with cooperative counterpart(s) in a region that has an environmental problem. These courses are structured to develop students' perception vis-á-vis environmental problems and to develop the skills required of environmental leaders. In addition, elective courses are chosen from degree programs, so that students can deepen both their professional knowledge and their skills. Research for master's and doctoral theses is also required and aims to propose new systems from interdisciplinary perspectives based on a holistic understanding of environmental issues.

1.3.1 Environmental Challenges and Leadership in Asia: Understanding Environmental Leadership

APIEL established a compulsory course, Environmental Challenges and Leadership in Asia, which is designed for students to acquire literacy and other skills required of an environmental leaders using a hands-on approach so that they can contribute to solving environmental issues or to build a sustainable society, which itself can prevent environmental problems. Students learn about real-world environmental problems in Asia from various perspectives, while considering the historical and cultural backgrounds. The course emphasizes group work and discussion, and students learn communication and consensus-building skills through experience. It leads to the award of two credits.

Over 15 lectures, this course examines the environmental problems that Asia is experiencing and discusses how we can foster a growing society with finite resources and space. At the same time, it helps develop the environmental leadership and other skills needed for building a sustainable society by taking up case studies of Asian environmental issues. One example: students are required to bring academic articles and books that they feel are the most important for the construction of a sustainable future. By sharing these reading materials, students become aware that perceptions of a sustainable future can differ widely. In addition, discussions among students help hone capabilities when developing cooperative relationships based on mutual understanding. In a sense, students in this compulsory course learn how to learn from others. The details of the compulsory course, Environmental Challenges and Leadership in Asia, are outlined in Chap. 3.

1.3.2 *Field Exercises: Developing Essential Skills for Environmental Field Sites*

APIEL has a strong focus on field exercises. The field exercises take place several times each year in cooperation with collaborating partners in Asia in loc.tions where there are environmental problems. A small number of students and faculty members form a group (a unit) and carry out hands-on exercise. Each field exercise setting is intended to broaden the perspectives of students and to cultivate an on-the-ground ability to identify and resolve problems in several ways, including pre-visit study and preparation, fieldwork, on-site experiments, discussions with stakeholders from many backgrounds including local educational institutions or governmental officials, a final presentation and report writing to solidify the vision and develop practical skills. Participants stay in the field for 1–2 weeks, with students receiving financial support for travel expenses. Overseas and domestic exercises lead to the award of two credits and one credit, respectively.

Tables 1.1, 1.2, 1.3, and 1.4 provide a list of the field exercises implemented to date. All field exercises have the following characteristics; they are (1) interdisciplinary in nature, (2) involvement of a number of different stakeholders, (3) fostering student initiatives, and (4) featuring practical issues that lack a prepared solution. Although the field study takes place over a relatively short period, it is possible for students to examine real-world environmental issues on-site and discuss them with the stakeholders. This interaction is made possible through the participation of local governmental agencies and companies and through the cooperation of collaborating universities/research institutions. Furthermore, this educational program is designed to ensure the diversity of participants (i.e., to avoid an overconcentration of Japanese graduates). This design feature then pushes students to improve their ability to communicate with people from other cultural backgrounds.

Table 1.1 List of APIEL's field exercise units (updated version of Akiyama et al. [1]): academic year 2009

	Overseas field exercise						Domestic field exercise		
Unit title	Intensive Program on Sustainability (IPoS)	Zhangye unit	Bangkok unit	Thailand unit	Chiang Rai unit	Greater Pearl River Delta (GPRD) unit	Eco-Industrial Cluster	Nissan-IPoS	Internship at Japan International Cooperation Agency (JICA)
Theme	Food, Energy and Water	Water-related issues in arid region	Urban development and agriculture-related issues in suburban Bangkok	Sustainable solid waste management in Asian developing countries	Transboundary environmental issues	Environmental LeadershipDevelopment in GPRD, China	Eco-Industrial Cluster	Sustainable mobility with zero emission vehicles	Sewage works engineering and Stormwater Drainage Technology
Place	Rayong Province, Thailand	Zhangye, Gansu Province, China	Bangkok, Thailand	Nonthaburi Province and Bangkok, Thailand	Golden Triangle, Chiang Rai, Thailand	Hong Kong and Guangzhou, China	Asian Development Bank Institute (ADBI) (Tokyo)	Yokosuka City, Kanagawa Prefecture	JICA (Tokyo)
Period	Aug. 1–12, 2009	Aug. 6–15, 2009	Sep. 14–23, 2009	Oct. 21–30, 2009	Dec. 19–30, 2009	Feb. 25–Mar. 7, 2010	Oct. 2009–Jan. 2010	Dec. 7–14, 2009	Nov. to Dec. 2009
Collaborator(s)	1. Asian Institute of Technology (AIT) etc.	1. Cold & Arid Regions Environment & Engineering Research Institute, Chinese Academy of Sciences (CAREERI), 2. Zhangye Water Authority	1. Chulalongkorn Univ. (CU)	1. AIT, 2. Kasetsart Univ. (KU)	1. Mae Fah Luang Univ. (MFLU)	1. The Hong Kong Univ. of Science & Technology (HKUST), 2. Sun Yat-sen Univ (SYSU)	1. ADBI	1. AIT etc.	1. JICA, 2. Sewerage Business Management Centre
Teaching staff(s)	Univ. of Tokyo (UT) (3); AIT (2)	UT (4); CAREERI (2)	UT (1); CU (2); Wakayama U (1)	UT (5); AIT (1); KU (1)	UT (3); MFLU (2)	UT (5); HKUST (1); SYSU (1)	UT (1)	UT (7); AIT (2)	UT (1)
Students' nationality	1 Ethiopian 2 Japanese N.B. 23 participants from 14 countries/ regions, 11 universities	1 Indonesian 1 Thai 1 Chinese 2 Japanese	1 Columbian 1 Filipino 2 Japanese	1 Filipino 1 Bolivian 3 Japanese	1 Indonesian 1 Cambodian 2 Sri Lankan 3 Japanese	1 Irish 1 Swiss 1 Bangladeshi 1 Portuguese 2 Chinese	1 Portuguese	1 Ethiopian 2 Japanese N.B. 24 participants from 15 countries/ regions, 11 universities	1 Brazilian 1 Chinese 1 Japanese

Table 1.2 List of APIEL's field exercise units (updated version of Akiyama et al. [1]): academic year 2010

	Overseas field exercise			Domestic field exercise				
Unit title	Oasis unit	Hue unit	GPRD unit	Coca-Cola Young Environmental Leaders Summit	Internship in the project on Low Carbon Green Asia	Tokyo Fringe unit	Nissan-IPoS	Green Energy unit
Theme	Sustainable integrated watershed management in cold and arid region	Flood and History in world heritage Hue city	Sustainable urban regeneration and relocation of industrial regions in GPRD, China	Corporate Social Responsibility	Low-carbon society Scenario in Asia	Urban development and agriculture-related issues in suburban Tokyo	Sustainable cities & mobility in 2050	Sustainable energy supply
Place	Zhangye, Gansu Province & Ejina, Inner Mongolia, China	Hue, Vietnam	Hong Kong and Guangzhou, China	Kuriyama Town, Hokkaido	ADBI (Tokyo)	Kashiwa, Chiba & Nerima, Kokubunji, Hachioji, Tokyo	Hayama town, Kanagawa Prefecture	Kashiwazaki, Niigata & Ueno, Gunma
Period	Aug. 10–23, 2010	Aug. 11–19, 2010	Feb. 21–28, 2011	Aug. 19–23, 2010	Nov. 2010 to Feb. 2011	Sep. 5–12, 2010	Dec. 4–13, 2010	Feb. 21–23, 2011
Collaborator(s)	1. CAREERI, 2. Zhangye Water Authority, 3. Alashan SEE Ecological Association, 4. Wusuronggui Village & Jirigalangtu Village, Ejina	1. Hue Univ.	1. HKUST, 2. SYSU	1. All counterparts of the overseas field exercise units in academic year 2009 & 2010, 2. Hokkaido Univ. (HU), 3. NPO Kuriyama, 4. Sumitomo Chemical Company Limited	1. ADBI	1. CU	1. AIT etc.	1. Tokyo Electric Power Company, 2. Korea Environmental Policy and Administration (KEPA) Society, 3. Prince of Songkla Univ. (PSU), 4. Kashiwazaki City

Teaching staff(s)	UT (5); CAREERI (3); Univ. of Niigata Pref. (UNP) (1)	UT (4); Hue Univ (1)	UT (4); HKUST (1); SYSU (1)	UT (4)	UT (2)	UT (1); CU (2); Wakayama U (1)	UT (7); AIT (2)	UT (5); KEPA (1); PSU (1); UNP (1)
Students' nationality	1 Ethiopian 1 Bhutanese 1 French 1 Vietnamese 1 Mongolian 1 Chinese 2 Japanese	1 Dominican 1 Bangladeshi 2 Filipino 1 Korean 1 Chinese 4 Japanese	1 Ethiopian 1 Australian 1 Filipino 1 Vietnamese 1 Malagasy 2 Chinese 2 Japanese	1 Irish 1 Indonesian 1 Ethiopian 1 Cambodian 1 Thai 1 Bangladeshi 2 Filipino 1 Bolivian 1 Portuguese 2 Japanese	1 Mongolian	1 Cambodian 1 Filipino 2 Japanese	1 Australian 1 Dominican 1 Chinese 1 Japanese N.B. 28 participants from 17 countries/regions, 10 universities	1 Irish 3 Indian 1 Indonesian 1 Cambodian 1 Thai 1 Dominican 1 Nepali 1 Bahraini 1 Bhutanese 1 Vietnamese 1 Jamaican 1 Chinese 1 Japanese

Table 1.3 List of APIEL's field exercise units (updated version of Akiyama et al. [1]): academic year 2011

	Overseas field exercise					Domestic field exercise			
Unit title	IPoS	Oasis unit	Thailand unit	Cambodia unit	GPRD unit	Ohtsuchi unit	Minamata unit	Nissan-IPoS	Coca-Cola Young Environmental Leaders Summit
Theme	Sustainable livelihoods through integrative practices with emphasis on Food, Water and Sanitation in a peri-urban community	Water-related issues in arid region	Sustainable Urban Water Use and Management in Bangkok	Sustainable Development of Cambodia	Sustainable urban development toward Green City: GPRD, China	Reconstruction of 3.11 Great Earthquake	Sustainable stakeholder management for water environment	Technology & Society - case of energy & transportation in Kashiwanoha	Corporate Social Responsibility
Place	Rayong Province, Thailand	Zhangye, Gansu Province, China	Bangkok, Thailand	Phnom Penh and Siem Reap, Cambodia	Hong Kong and Guangzhou, China	Ohtsuchi Town and Kamaishi City, Iwate Prefecture	Kumamoto City and Minamata City, Kumamoto Prefecture	Kashiwa City, Chiba Prefecture	Kuriyama Town, Hokkaido
Period	Aug. 1–12, 2011	Aug. 27–Sep. 7, 2011	Aug. 18–28, 2011	Sep. 2–14, 2011	Feb. 15–25, 2012	11 times of 3–4 day fieldwork	Jan. 4–8, 2012	Dec. 5–12, 2011	Feb. 27–Mar. 3, 2012
Collaborator(s)	1. AIT etc.	1. CAREERI, 2. Zhangye Water Authority	1. AIT, 2. KU	1. Royal University of Phnom Penh (RUPP), 2. Seoul National University (SNU), 3. JICA, 4. Korea International Cooperation Agency (KOICA)	1. HKUST, 2. SYSU, 3. SNU, 4. Univ. of Hong Kong (UHK), 5. Hong Kong Green Building Council (HKGBC) 6. Guanzhou Urban Land Administration Bureau (GULAB)	1. Ohtsuchi town 2. Coastal Regional Development Bureau of Iwate Prefecture	1. Kyushu Univ., 2. Kumamoto Univ., 3. Univ. of Kitakyushu, 4. Univ. of Tsukuba	1. AIT etc.	1. All counterparts of the overseas field exercise units in academic year 2011, 2. HU, 3. Yokohama National Univ. (YNU), 4. JICA

	UT (9); AIT (2)	UT (3); Public Works Research Inst. (1); CAREERI (3); UNP (1)	UT (5); AIT (1); KU (1)	UT (4); RUPP (2); SNU (2)	UT (5); HKUST (2); SYSU (3); SNU (2); UHK (1); HKGBC (1); GULAB (1)	UT (9)	UT (4); Kyushu U (3); Kumamoto U (4)	UT (5)	UT (6), HU (2), YNU (1), JICA (1)
Teaching staff(s)									
Students' nationality	1 Chinese 2 Japanese N.B. 22 participants from 10 countries/ regions, 9 universities	1 Brazilian 2 Cambodian 1 Korean 1 Japanese 1 Sri Lankan	2 Chinese 1 Indian 1 Nepalese	1 Chinese 1 Indian 4 Japanese 1 Nepalese	1 Bangladeshi 1 Chinese 1 Indian 2 Japanese 1 Korean	2 Chinese 7 Japanese 1 Mongolia 1 Sri Lankan	1 Chinese 1 Japanese 1 Indian 1 Nepalese	1 Chinese 2 Japanese N.B. 18 participants from 9 countries/ regions, 8 universities	1 American 2 Cambodian 8 Chinese 2 Indian 3 Japanese 2 Korean 3 Nepalese 1 Mongolia 2 Myanmar 1 Sri Lankan 3 Thai 2 Vietnamese

Table 1.4 List of APIEL's field exercise units (updated version of Akiyama et al. [1]): academic year 2012

	Overseas field exercise						Domestic field exercise		
Unit title	IPoS	Oasis unit	Thailand unit	Cambodia unit	GPRD unit	Bangladesh unit	Nissan-IPoS	Minamata unit	Coca-Cola Young Environmental Leaders Summit
Theme	Livelihood strategies for adaptation to climate change	Water-related issues in arid region	Sustainable Urban Water Management: special focus on flood management	Sustainable Development in Cambodia	Sustainable urban development in GPRD, China	Risk assessment through food and water in rural community in Bangladesh	Climate Change & Society—case of energy issues in Kashiwanoha	Role of scientists, policy makers and citizens: Case of long lasting Minamata Disease	Corporate Social Responsibility
Place	Nakornnayok and Pathum Thani, Thailand	Zhangye, Gansu Province, China	Bangkok & Ayutthaya, Thailand	Phnom Penh and Siem Reap, Cambodia	Hong Kong and Guangzhou, China	Dhaka, Manikganji & Comilla, Bangladesh	Kashiwa City, Chiba Prefecture	Minamata City, Kumamoto Prefecture	Kuriyama Town, Hokkaido
Period	Jul. 28–Aug. 8, 2012	Aug. 4–16, 2012	Aug. 19–27, 2012	Aug. 4–14, 2012	Feb. 21–Mar. 3, 2013	Mar. 1–10, 2013	Dec. 10–16, 2012	Nov. 21–25, 2012	Feb. 17–23, 2013
Collaborator(s)	1. AIT etc.	1. CAREERI, 2. Zhangye Water Authority, 3. Sophia Univ.	1. AIT	1. RUPP, 2. SNU, 3. JICA, 4. KOICA	1. HKUST, 2. SYSU, 3. SNU, 4. UHK, 5. Chinese Univ. of Hong Kong (CUHK), 6. HKGBC 7. GULAB	1. Dhaka University of Engineering & Technology (DUET)	AIT etc.	1. Kyushu Univ., 2. Kumamoto Univ., 3. Univ. of Kitakyushu, 4. Univ. of Tsukuba	1. All counterparts of the overseas field exercise units in academic year 2012, 2. HU
Teaching staff(s)	UT (7); AIT (6); Phranakhon Rajabhat U (1); Srinakharinwirot U (1); Thammasat U (1)	UT (3); CAREERI (3); Sophia U (1); UNP (1)	UT (5); AIT (2)	UT (4); RUPP (2); SNU (2)	UT (5); HKUST (2); SYSU (3); SNU (2); UHK (1); CUHK (1); HKGBC (1); GULAB (1)	UT; DUET	UT (4); AIT (4); Srinakharinwirot U (1); Chiangmai U (1)	UT; Kyushu U; U of Tsukuba; Kumamoto U; U of Kita-kyushu	UT (4)

Students' nationality									
	1 Chinese 1 Sri Lankan 1 Japanese N.B. 21 participants from 16 countries/ regions, 8 universities	1 American 1 Bangladeshi 3 Chinese 1 French 1 Japanese 1 Korean	1 Brazilian 2 Filipino 3 Japanese 1 Korean 1 Sri Lankan 1 Thai	1 Cambodian 1 Indonesian 1 Japanese 1 Korean 1 Swiss 1 Swedish	1 Bahraini 1 Colombian 1 Indonesian 2 Japanese 1 Jordanian 1 Malagasy	1 Chinese 1 French 1 Japanese	1 Chinese 1 Sri Lankan N.B. 19 participants from 15 countries/ regions, 8 universities	2 Chinese 1 Colombian 1 Filipino 1 French 1 Ghanaian 1 Indonesian 1 Japanese 1 Korean 1 Thai	TBD

APIEL also organized activities that have horizontal links with different field exercise units, including joint presentation meetings that involve a variety of field exercise units, the Coca-Cola Young Environmental Leaders Summit Unit (Tables 1.1, 1.2, 1.3, and 1.4), as well as student sessions in workshops and international symposia that help deepen discussions with domestic and international experts. Through these joint activities, it is possible for participants to share not only the lessons learned from their own fieldwork, but also to learn other approaches to several types of environmental issues. Teaching staff, in particular, join most final presentation meetings, participate in field exercise units, and further develop their own educational skills. In this way, APIEL is also directing its energies into developing new educational methods that will nurture a more comprehensive range of human resources. The details of compulsory field exercises are covered in Chaps. 4–7 (Thailand Unit, Oasis Unit, GPRD Unit, and Cambodia Unit).

1.3.3 Elective Courses: Enabling Interdisciplinary and Specialized Approaches

APIEL offers a group of elective courses based on the specialty of GPSS and UE. This has the advantages of both the interdisciplinary approach toward sustainability science provided by GPSS and the specialized knowledge of environmental engineering provided by UE. Students learn about a broad range of environmental issues in Asia regardless of their major field of study. All APIEL elective courses are offered in English. Although each course is single-discipline oriented, the overall selection of a group of courses takes interdisciplinary elements into account. To date, the following courses have been offered by GPSS and UE:

1.3.3.1 Graduate Program in Sustainability Science (GPSS)

Sustainability Perspectives in Environmental Issues; Strategies for Global Sustainability; Introduction to Natural Environmental Studies; Residential Environment; Environmental Economics, Environmental Business; Business and Finance for Sustainable Development; Innovation and Sustainability; Environmental Sustainability; Urban Sustainability in Relation to the Water Sector; Sustainable Health and Environment; Sustainability Education; Frontier of Sustainability Science; Environmental Sustainability, Concepts and Methodologies of Sustainability Science; Planning and Design and for Sustainability; Advanced Concepts and Methodologies of Sustainability Science; Management and Policy Studies of Sustainability; Sustainability of Resources; Planning and Design for Sustainability.

1.3.3.2   Department of Urban Engineering (UE)

Management of Global and Urban Environments; Water Environment Technology; Urban Water Systems; Fundamentals of Water Pollution Control; Solid Waste Management; Environmental Risk Management; Hazardous Waste Management; Regional Planning; Urban Development Policy and Planning; Urban Transport Policy; Urban Transport Planning and Analysis; Environmental Sanitation in Developing Countries; Urban Planning in Developing Countries; Advanced Water Quality Engineering; Advanced Wastewater Engineering; Systems and Tools toward the Recycle-based Society; Advanced course in Health-related Water Microbiology.

1.4   Requirements for Completing the Program

Figure 1.3 shows the requirements for completion of the program. Students satisfying the completion requirements for APIEL are presented with The University of Tokyo Certificate of Completion for Asian Program for Incubation of Environmental Leaders. Requirements for completing the program are (1) to complete the APIEL's compulsory course of "Environmental Challenges and Leadership in Asia" (earning

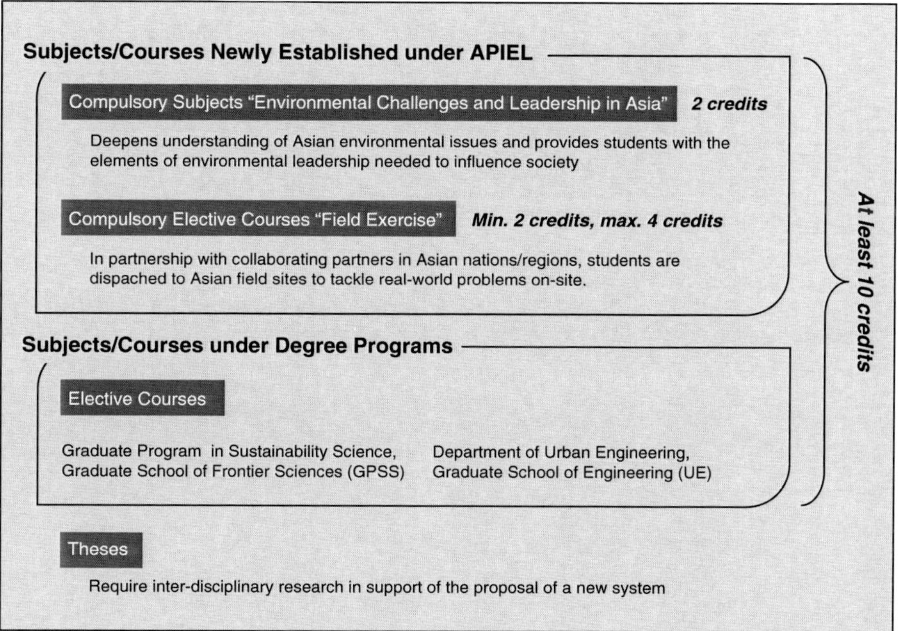

Fig. 1.3 Requirements for completing the program

two credits) and compulsory elective courses from the field exercises (minimum two credits, maximum four credits), (2) to earn at least ten credits in total from compulsory courses, compulsory elective courses, and other elective courses, and (3) to complete a postgraduate degree program either in GPSS or UE.

Reference

1. Akiyama T, Li J, Onuki M (2012) Integral leadership education for sustainable development. J Integral Theory Pract 7:72–86

Chapter 2
The Concept of Environmental Leader

Tomohiro Akiyama, Kyoungjin J. An, Hiroaki Furumai,
and Hiroyuki Katayama*

Abstract Leadership has shifted focus from the individual to the group or institute. Efforts to link leadership and the natural environment have already begun and the necessity for environmental leadership has never been higher than ever in the era of complex and evident environmental and social problems, such as climate change, global conflict, limited resources, an overwhelming amount information, etc. There is no single solution for environmental problems that can solve the conflicts of diversified community relations. Therefore, environmental leadership development is a priority element for improving the deteriorating environment. However, the current education system, especially in Asia, lacking in providing a holistic view of environmental issues, as well as inter- or trans-disciplinary and cross-cultural approaches, or a balance between the environmental, economic and social dimensions, using hands-on experience.

*All the authors contributed equally to the article and are listed alphabetically.

T. Akiyama (✉)
Graduate Program in Sustainability Science, Graduate School of Frontier Sciences,
The University of Tokyo, Environmental Studies Building 334, 5-1-5 Kashiwanoha,
Kashiwa, Chiba 277-8563, Japan
e-mail: akiyama@k.u-tokyo.ac.jp

K.J. An
Asian Program for Incubation of Environmental Leaders (APIEL), Department of Urban
Engineering, Graduate School of Engineering, The University of Tokyo,
7-3-1 Hongo, Bunkyo-ku, Tokyo 113-8656, Japan
e-mail: ankj@env.t.u-tokyo.ac.jp

H. Furumai
Research Center for Water Environment Technology, Graduate School of Engineering,
The University of Tokyo, 7-3-1 Hongo, Bunkyo-ku, Tokyo 113-8656, Japan
e-mail: furumai@env.t.u-tokyo.ac.jp

H. Katayama
Department of Urban Engineering, Graduate School of Engineering,
The University of Tokyo, 7-3-1 Hongo, Bunkyo-ku, Tokyo 113-8656, Japan
e-mail: katayama@env.t.u-tokyo.ac.jp

T. Mino and K. Hanaki (eds.), *Environmental Leadership Capacity
Building in Higher Education*, DOI 10.1007/978-4-431-54340-4_2,
© The Author(s) 2013

In response, APIEL strives to fill this gap by improving education for environmental leadership with sustainability issues in mind. This chapter will review the concept of the environmental leader through a discourse on leadership. As well, it will introduce the authors' experiences in fostering environmental leader by establishing and implementing environmental leadership education over the past four years. The discourse on environmental leadership illustrates how environmental leaders have been educated to cope with emerging environmental issues. The concepts of transformational/transformative-, eco-, collective, green, and communicative leadership provide a map to understand the evolution of the theory and practice of environmental leadership education.

Keywords Discourse on leadership • Environmental issues • Environmental leadership • Leadership experience

2.1 Discourse on Environmental Leadership[1]

What is so different about environmental leadership from leadership in other areas? To answer this question, this chapter begins with the fact that very little research has been done on the issue of environmental leadership, although a comprehensive review of the literature on leadership itself uncovers focuses on business, political, and public leadership. Books on business leadership typically contain neither a substantive analysis of the psychology of future orientation nor a sense of the larger systemic constraints on future activities that must be taken into account by leaders [1]. For instance, the idea of a leader for the future contains little reflection on the larger systemic constraint on future activity, including climate change, as Heifetz [2] put it, "our focus on the production of wealth rather than coexistence with nature has led us to neglect fragile factors in our ecosystem." However, recently, scholarly textbooks on leadership have addressed the importance of the natural environment as a significant context for leadership, or as an "emerging issue" of interest [3], and Heifetz [2]'s theory of adaptive leadership provides an important starting point for thinking about environmental leadership.

2.1.1 History of Environmental Education and the Need for Environmental Leaders

Environmental education has traditionally focused on how to foster changes in individuals that are associated with pro-environmental actions and behaviors [4]. As Table 2.1 shows, environmental education has developed over the past 50 years from the perspectives of natural resource management or the management of environmental organizations. Few researchers have placed the relationship between leadership and the natural environment at center stage and examined it from diverse viewpoints [5].

---

[1]This section is written by one of the authors, Kyoungjin J. An from Department of Urban Engineering, Graduate School of Engineering, The University of Tokyo, Japan.

Table 2.1 History and trend of environmental education (adapted from Palmer [6])

	Environmental education: key events on a development timeline	Key trends in environmental education and leadership development
	1948 The International Union for Conservation of Nature (IUCN) conference; first use of the term environmental education **1949** Founding of IUCN	
1960s ↓	**1968** The United Nations Educational, Scientific and Cultural Organization (UNESCO) Biosphere Conference, Paris	**Nature study** Learning about plants, animals and physical systems **Fieldwork** Led by an "expert" with a particular academic focus: biology, geography, etc.
1970s ↓	**1972** UN Conference on the Human Environment, Stockholm **1975** Founding of the United Nations Environment Programme (UNEP) and Institute for European Environmental Policy (IEEP) UNESCO/UNEP International Workshop on Environmental Education, Belgrade **1977** UNESCO First Inter-Governmental Conference on Environmental Education, Tbilisi	**Outdoor/adventure education** Increasing use of the natural environment for first-hand experience **Field studies centers** Growth of field and environmental/outdoor education for developing awareness through practical activity and investigation **Conservation education** Teaching about conservation issues **Urban studies** Study of the built environment, street work
1980s ↓	**1980** World Conservation Strategy (IUCN, UNEP, World Wildlife Fund (WWF)) **1987** UNESCO/UNEP Educational Congress on Environmental Education and Training, Moscow World Commission on Environment and Development—Our Common Future—The Brundtland Report	**Global education** A wider vision of environmental issues **Development education** Environmental education has a political dimension **Value education** Clarifying values through personal experience **Action research** Community problem solving, student-led problem solving involving fieldwork
1990s ↓	**1990** Publication of *Caring for the Earth: A Strategy for Sustainable Living* (IUCN et al.) **1992** UN Conference on Environment and Development –"The Earth Summit"	**Empowerment** Communication, capacity building, problem solving and action, aimed at the resolution of socio-environmental problems **Education for a sustainable future** Participatory action, relevant approaches to changing behaviors and resolving ecological problems
2000s		**Community** Students, teachers, NGOs, and politicians working together to identify and resolve socio-ecological problems

A comprehensive list of readings and reports on environmental education touches on every aspect of sustainability as we learned from Table 2.1, but the link with leadership angle is left unexplored till the early 1990s. Gunderson et al. [7] provide insights into leaders and managers trying to solve natural resource and other environmental problems. Without mentioning leadership, Moser and Dilling [8] provide a comprehensive look at the communication challenges presented by climate change. In the twenty-first century, environmental education is turning toward a community of partners so that students, teachers, NGOs, and politicians can work together to identify and resolve socio-ecological problems. In order to make these fundamental changes, leadership is pivotal for driving the change. As well, leadership capacity is needed across a broad range of stakeholder groups, including politicians, city officials, and emergent leaders in the public and private sector, also researchers, educators, communities, and individual citizens [9].

Today, the challenge for fostering leaders is how higher education policy and provisions can be reoriented or retrofitted in a way that is organizationally practicable, academically acceptable, and educationally sound. Higher educational institutions have a crucial role to play. Therefore, when we say *reorienting higher education*, the tasks we have to observe are how the existing knowledge on environmental education and sustainability can be extended as well as establishing the role that the higher educational institutions can play. There must be a balance in society between investing for sharing existing scientific knowledge and further extension of that knowledge. Our strong emphasis on leadership pedagogy reflects our belief that under the present circumstances it is more important to extend basic knowledge of how the world works for the common good than for a few specialists to master further details of their special disciplines. As described in Chap. 1, APIEL was born through such reoriented environmental education at The University of Tokyo. However, Asian universities have been relatively slow or distant from the global movements of networking in environmental education and leadership development. In addition, participatory leadership programs in higher education in Asia have been weak so far while various international environmental leadership programs have been launched by the United Nations Environment Programme and/or universities in Europe and North America.

2.1.2 Evolution of Environmental Leadership Over Time and Space

Leadership in this chapter entails environmental concerns and takes place in two dimensions: space and time. The type of leadership—transactional leadership shown by the Western industrial paradigm of the past few 100 years—premised on using resources regardless of social and environmental concerns is no longer desirable in a world that recognizes universal human rights and sustainable development. The contemporary view of leadership can be defined as a process of influence that occurs within the context of relationships between leaders and followers, and involves establishing vision, aligning resources, and providing inspiration to achieve

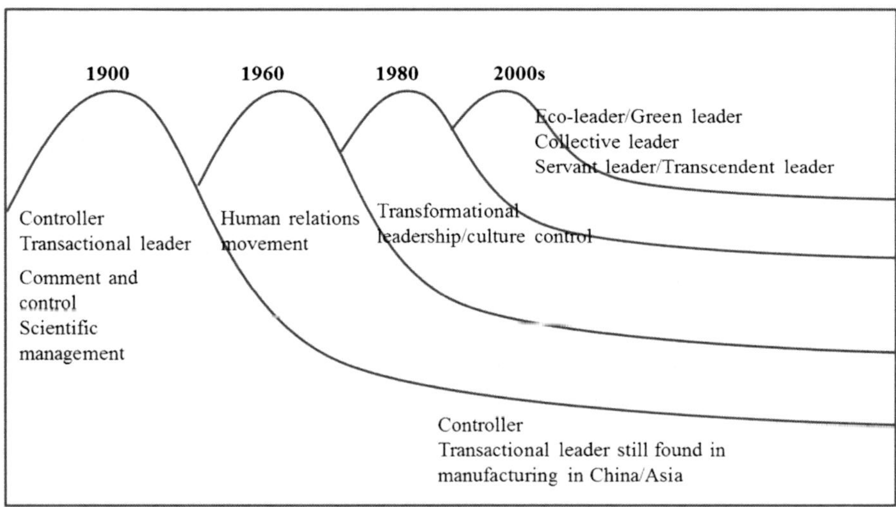

Fig. 2.1 Discourse on leadership. Adopted from Western [13]

mutual interest. Although, there are many leadership theories, none are universally accepted in environmental leadership. However, transformational leadership has often been featured strongly in studies of environmental leaders [10] and this theory is helpful in understanding and explaining the behaviors of environmental leaders. For instance, Burns [11] defines leadership as transformational leaders inducing followers to act for certain goals that represent the values and the motivations of both leaders and followers. Greenleaf [12]'s servant leader is more explicitly visionary than the transforming leader, in that leaders must also have a prophetic vision of the future state into which followers are being led. Collective leadership that includes cultural and social perspectives to deal with complex environmental issues is addressed in following Sect. 2.2. We depend on leaders to respond to time and place in a situation of diminishing natural resources and growing environmental degradation. Figure 2.1 illustrates the discourse on leadership and how environmental leadership has evolved spatially and over time in response to needs [13].

2.1.3   Becoming an Environmental Leader

Norman L. Christensen Jr. has written his perspective on environmental leadership in the foreword to the book *Environmental Leadership Equals Essential Leadership* [14]. He states that there are four ingredients for environmental leadership: boundaries, priorities, uncertainties, and action [14]. Among the most interesting and daunting characteristics of environmental issues is addressed: the extent to which they permeate virtually all *boundaries*. For instance, decisions to cut or not cut a forest in one region have an effect on the nature of forest management in other regions, and the solutions to such environmental problems demands communication, understanding, and collaboration among diverse disciplines and traditions. Berry and Gordon [5] argue convincingly that the antidote to the boundary problem is problem-oriented, systems-based thinking. Effective environmental leaders assess the extent of a challenge by the spatial and temporal scale of a physical or biological process, as well as its cultural, social, and institutional elements, rather than trying to destroy or redefine boundaries. Environmental challenges appear to be limitless, while the resources used to meet these challenges are limited. Therefore, leaders must be able to set clear *priorities* among the needs and demands. *Uncertainty* is a major challenge for leaders in many circumstances, but most especially in those related to an environment of ignorance, variability and complexity. If leaders are certain beforehand, they can lead more effectively, but the truth is that environmental leaders must constantly act in the context of uncertainty and change. The connection between leadership and *action* is described as change. Successful leadership depends on a clearly stated vision and goals as well as models that connect the action that drives change.

Berry and Gordon [5] stated that leadership, at least in terms of environmental leadership, is not yet sufficiently congruent with any theory for it to provide a reliable basis for thought and action. In this view, experience, observation, and individual thinking must substitute for a theory as the basis for teaching. For classroom education, they included four elements: (1) a vision on the characteristics of leaders, which is produced from the practice of thinking ahead using all available data, and testing predictions and insights, and thus it becomes learnable and teachable; (2) leadership skills, such as ethics and personal values, communication, management, conflict assessment and resolution, influencing legislation and policy, and fundraising; (3) an observation of leaders themselves, knowing that effective environmental leadership very much depends on the context, e.g., organizational culture, geographic location, variability in the natural environment, etc.; and (4) the construction of leadership prescriptions for real organizations and real situations, which includes the current status, challenges and problems, as well as options for the solutions, and indicators to monitor and assess the solutions. The traditional model of a hierarchical leader with strong authority was replaced by the leader who worked in a participatory team environment where goals were created in a collaborative and shared decision-making process.

Environmental leaders who promote environmental sustainability infuse their desire to protect the natural environment into their decision-making and action processes [15]. Although the traditional focus of environmental leaders has been on individual development, today it involves two levels of influence: individual and institutional. So, in this regard, leadership can be found as both individual and group-based change. This leadership skill is more likely to be successful if leaders understand how to influence interrelated processes at all levels. However, the process of converting one individual at a time is slow and is unlikely to accomplish major change quickly, unless there are highly visible indicators of progress in the form of policies, programs, and budgets. Therefore, leadership development is a process of creating change agents in society, which is a complex psychological and social process. Describing the change process for individuals participating in environmental leadership development programs may be as complex and challenging as describing the change process in future transformational leaders. Leadership development programs that aim to build this transformational leadership capacity are rare in an Asian context. This chapter then argues that building leadership capacity within higher academia is a way educators can work to affect desirable behavioral change and advance sustainable environmental management practices. The following sections (2.2–2.4) will introduce other author's concepts of environmental leaders from their experience applied to APIEL as educators in higher academia. As such, collective leadership, institutional leadership and strong leadership under disaster management are introduced.

2.2 Exercising Collective Leadership to Find Solutions for Global Environmental Issues[2]

This section outlines the author's experience of conducting research in the Heihe River basin in northwestern China. From this and other experiences, it is apparent that the complexity of environmental problems requires collective leadership that includes cultural and social perspectives. One possible definition for collective leadership, following Akiyama et al. [16], is also included.

2.2.1 Introduction

It is widely acknowledged that environmental problems are the problems of complexity. One of the dimensions of complexity arises from conflicts of the vested interests of the various stakeholders. However, how do you balance the interests of stakeholders with sustainable development that includes environmental concerns? Following the author's own experiences in the Heihe River basin in northwestern China, he feels that collective leadership might be a key in reaching a consensus among stakeholders that ultimately leads to the shared visions/goals for solving global environmental problems. The author will outline some of his experiences through participating in an interdisciplinary research project called Historical Evolution of the Adaptability in an Oasis Region to Water Resource Changes, carried out from academic year 2001 to 2006.[3]

This section is based on the integral framework for environmental leadership education proposed in Akiyama et al. [16]. Figure 2.2 shows the framework that was

---

[2]This section is based on the personal experiences of one of the authors, Tomohiro Akiyama from Graduate Program in Sustainability Science, Graduate School of Frontier Sciences, The University of Tokyo, Japan.

[3]Further information about the project can be found at their website (http://www.chikyu.ac.jp/rihn_e/project/H-01.html).

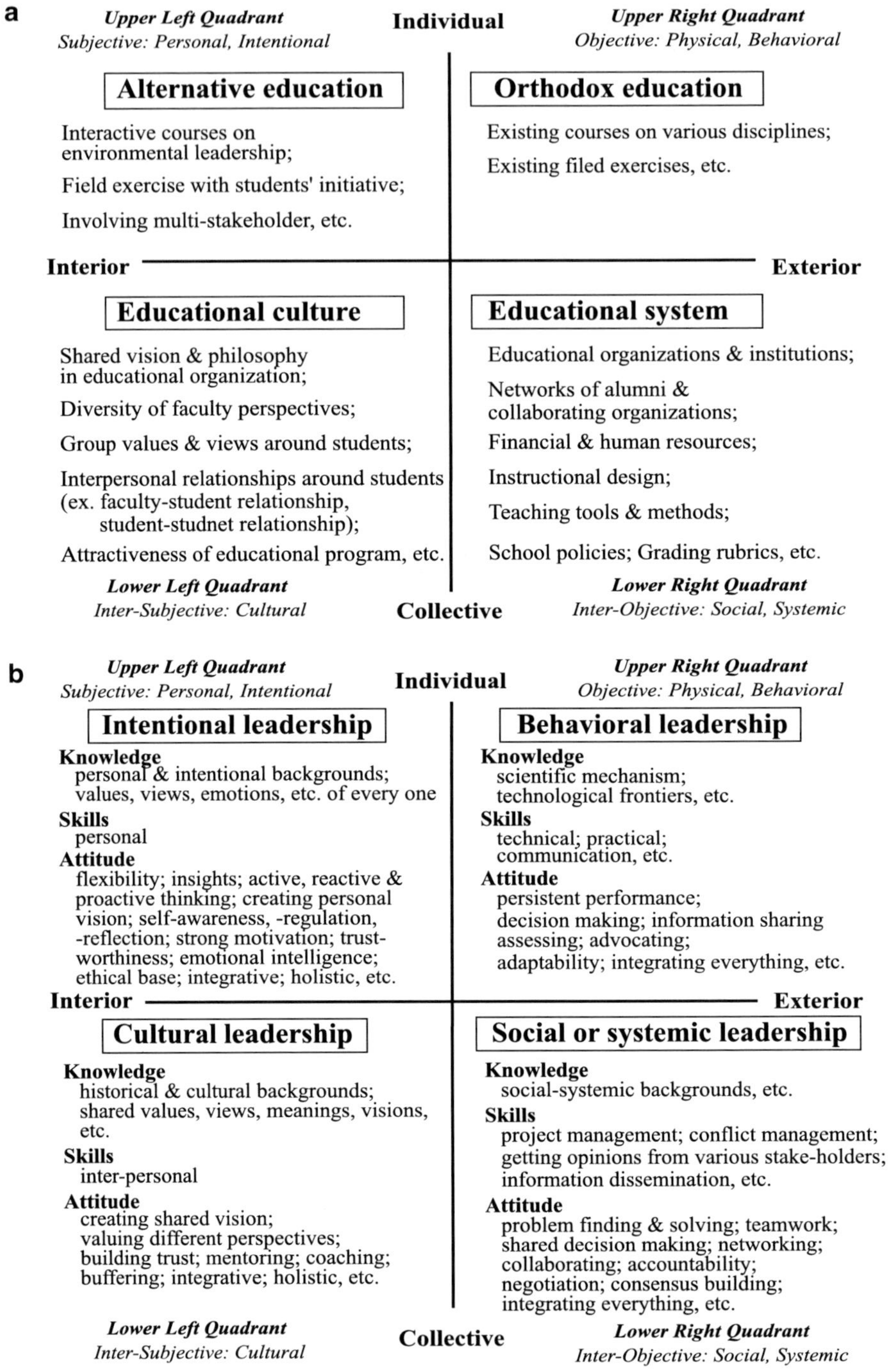

Fig. 2.2 Four-quadrant model of integral leadership education for sustainable development [16]

developed by modifying Integral Approach of Wilber [17, 18].[4] The upper quadrant in Fig. 2.2a shows the components of education programs, while Fig. 2.2b presents components of environmental leadership inherent in students. In this framework, collective leadership is considered to incorporate both cultural leadership and social leadership within leadership theory.[5] This section considers that collective leadership is the key to building general environmental leadership skills.

2.2.2 Experiences in the Heihe River Basin

Water is essential for life. However, water environmental problems, including the drying up of rivers/lakes, declining groundwater tables, vegetation degradation, and desertification, are intensifying in many regions. The Heihe River basin in northwestern China is a good illustration of these problems on the scale of a river basin. The Heihe River is the second largest inland river in China. Its basin area covers about 130,000 km², and it is 821 km long. Roughly speaking, this river flows through three administrative provinces in China: the upper reaches are in Qinghai Province, the middle reaches are in Gansu Province, and the lower reaches are in the Inner Mongolia Autonomous Region. In the river basin, people rely on glacier meltwater and precipitation from the mountainous areas in the upper reaches. Over the past 2,000 years, the Han and Mongolian people have lived in the middle reaches (oasis region) and lower reaches (desert region), respectively [20–22].[6] Due to these geographical differences, the Han mostly practiced settled farming, while the Mongolians were nomadic pastoralists (herdsmen). Throughout their history, the people living in the middle reaches and the people living in the lower reaches had continual conflicts over water. The conflict has dramatically worsened since the 1950s. The people living in the middle reaches, benefiting from technological improvements to dig deep wells with electric pumps, have been able to reclaim and irrigate large areas of farmland on the fringes of the oasis. Those lands do not have direct access to river water, so in the past, people were not able to grow crops there. However, given access to groundwater, the people have been expanding the area of irrigated farmland; their use of water intake for irrigation, as a consequence, has

---

[4]Wilber [17, 18] proposed an integrated method called the all-quadrants, all-levels (AQAL) model, which is gaining attention in the education field as an effective way of teaching and designing curricula [19]. Wilber [17, 18]'s integrated methodology features a four-quadrant framework which contends that reality is composed of holons. All holons have both an objective exterior expression and a subjective interior experience. At the same time, all holons are both individuals and members of a collective. These two distinctions between the exterior and interior, and the individual and collective, give rise to four aspects of reality, or four ways of knowing, represented by the quadrants. Although the four quadrants are ontologically distinct, Wilber [17, 18] suggests that there is an interwoven, intimate correspondence between all four quadrants.

[5]In the literature, there are many definitions of collective leadership.

[6]More ethnic groups live in the river basin. Han and Mongolian people are simply the majority groups in the middle and lower reaches, respectively.

Fig. 2.3 A river disappearing into the desert, lower reaches of the Heihe River basin (January 2002)

Fig. 2.4 Wells buried by sand (June 2002)

increased. As a result, the water intake in the middle reaches has increased substantially, while the river discharge to the lower reaches has declined (Figs. 2.3 and 2.4).

The author visited the Heihe River basin for the first time in 2001. That summer, as a graduate student majoring in hydrology, he joined a research project that was investigating the water balance of the river basin. He was in charge of the study of the lower reaches. For the research, the author travelled the desert to find Mongolian nomadic households and to investigate their wells. Though a total stranger, he was welcomed warmly with tea and wine. He sometimes spent nights in their yurts, and was served fresh goat meat, which was the best treat in the region. The hospitality was impressive. Every time he returns to the region, the author tries to visit the same

Fig. 2.5 Mongolian teaching Mandarin to his granddaughter (May 2004)

families. However, reunions with old friends have become difficult in recent years. Many families are not living on their pasture land anymore. This is not because they are moving seasonally, instead it is more related to the implementation of a government policy for environmental conservation (Fig. 2.5).

Environmental degradation in the Heihe River basin, mostly in the lower reaches, has attracted a lot of attention in China since the late 1990s. The two terminal lakes of the Heihe River have dried up, in 1961 and 1992, respectively. Due to mass media reports, the area around the dry lakes is believed to be the origin of sandstorms in northern China. As a result, the Chinese government promulgated the Integrated Water Resources Management Plan of the Heihe River basin in 2001. In the lower reaches, the detailed plan was implemented in 2002 to help in the recovery of the environment. The detailed plan includes putting fences around 20,000 ha of riparian forest, establishing 2,700 ha of farmland for growing fodder, digging 110 wells with electric pumps, and relocating 1,500 nomads from grasslands. The logic behind the policy for the lower reaches was that the nomadic style of raising livestock was the fundamental reason for the environmental degradation. Therefore, to promote "economically efficient" livestock raising, the government created incentives for nomads to raise animals in shelters. The government built houses with livestock shelters close to the center of the town, and provided subsidies to the nomads who moved into them. The wells and reclaimed farmland were dedicated to raising livestock in fixed locations.

Were these policies effective for environmental conservation? Mongolian nomads in the lower reaches generally do not think so. They said that although the government required the middle reaches to release a certain amount of water to the lower reaches every year, they only released water *after* irrigation period in the middle reaches. The nomads also said that now the grasslands cannot recover, and the water release simply brings a flood over a short period. In addition, the water was diverted to the terminal lakes through man-made concrete channels, so riparian vegetation cannot get enough water. Meanwhile, grassland without livestock is facing other problems, such as an overpopulation of mice.

These policies, which ignored the indigenous Mongolian culture and their expertise in raising livestock, in turn upset the balance of ecological systems in the lower reaches. On the other hand, in the middle reaches, the government restricted the intake of river water. However, no-one paid much attention to groundwater management. Han farmers, living in the middle reaches, simply changed from river water irrigation to groundwater irrigation. As a consequence, declining groundwater discharge into the river resulted in less river discharge in winter (non-irrigation period) in the lower reaches because of the complex interaction of the systems [23]. In short, the results from our research project clearly showed that the environmental degradation was caused by farmland expansion in the middle reaches, instead of the overpopulation of livestock in the lower reaches [24–28].

The key environmental problems in the Heihe River basin are how to effectively distribute water between the middle and lower reaches and how to support the different styles of subsistence as well as the natural environment. As we have learned, nomadic pastoralism, as an adaption to the spatial distribution and the temporal variability of arid regions, exists across the Mongolian steppes and Eurasia. It disperses the burden on the environment through movement across the grasslands. On the contrary, settled farming requires more water. The fodder growing promoted by the government actually created a new need for water, or in other words, a burden on the environment.

2.2.3   Collective Leadership for Finding Solutions for Environmental Problems: A Personal View

In recent years, environmental issues are attracting substantial attention from academics, policymakers, businesspeople, and the general public. However, policymakers and their think tanks and academic institutions are often outsiders who do not have an adequate understanding of indigenous cultures and other information about the target regions. Thus, the ideal image drawn by policymakers and researchers may not reflect the needs of local people. Advanced technologies may not be a good match for the real circumstances in target regions. Good intentions to solve environmental issues, as a result, can cause disagreement, or even conflicts.

In the Heihe River basin, the author strongly feels the need to build effective communications between the middle and the lower reaches—the Han farmers and the Mongolian nomads—as well as communications between policymakers and local people. Consensus building through discussions and mutual understanding might be the first step in solving this problem. In Fig. 2.6, the author's Mongolian friend, born in the lower reaches, and a Han friend, born in the upper reaches, are shown enjoying a party together after a day of hard work. The Han friend, due to his diligence and kindness, was called *Jaahandai* by the local Mongolian people, a word often used to describe something small but lovely. This nickname shows their hospitality and

Fig. 2.6 Party after a workday (September 2004)

friendship towards the young man. Going back to the start of this section, the author believes that collective leadership incorporating cultural and social perspectives is one of the keys to solving current global environmental problems.

2.3   Required and Expected Abilities and Skills for Environmental Leaders in Asia[7]

2.3.1   Environmental Problems and the Need for Environmental Leaders in Asia

With an increasing population and rapid economic growth in Asia, the demand for food and water resources has expanded. In addition, environmental pollution and health hazards are also becoming more pronounced. Therefore, there is an urgent need to deal with these problems. Without solutions to environmental problems in Asia, we might face a serious situation for sustainability on a global scale, as shown in Fig. 2.7.

However, human resources are required to resolve the issues of poverty, urbanization and industrialization, as well as environment problems. Therefore, the development of people who can contribute to the solution of emerging environmental problems is urgently needed in Asia. As well as responding to these problems, it is

[7]This section is based on the personal experiences of one of the authors, Hiroaki Furumai from Graduate School of Engineering, The University of Tokyo, Japan.

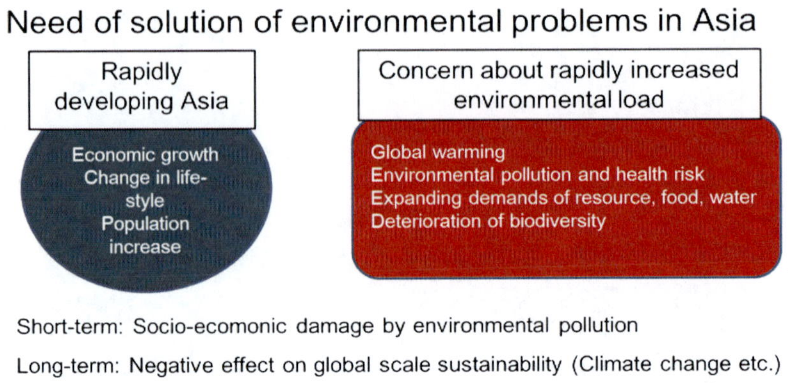

Fig. 2.7 Need of environmental leaders in Asia (http://www.env.go.jp/press/press.php?serial=9516)

necessary to realize a low-carbon, recycling-based society, one in harmony with nature from a long-term perspective. Toward the creation of a sustainable society, it is essential to have people who can internalize environmental protection and conservation into today's society.

It has been recognized that people are a part of nature in Asia. In addition, environmental ethics and wisdom from traditional philosophies of "Enough is as good as a feast" have been handed down over the generations. Recently, we have tended to forget the historical accumulated wisdom, while we put more weight on economic growth and efficiency. We have to re-recognize the importance of environmental ethics and wisdom. In addition, leaders with a long-term perspective are required to achieve a sustainable development in Asia.

2.3.2 Essential Elements of Environmental Leaders

Committee on the Vision for Developing Environmental Leaders in Higher Education towards Achieving a Sustainable Asia, the Ministry of Environment, Japan, reported/discussed the key concept/component/element of environmental leaders. They proposed that environmental leaders are required to have "strong motivation," "expertise" and "leadership," as shown in Fig. 2.8.

Strong motivation should be based on a clear understanding of the urgency of the current state of sustainability and action. Willingness to act for environmental protection is expected of environmental leaders. Environmental ethics and the ability to assess long-term and short-term needs are closely related to the strong motivation of environmental leaders. At the same time, motivation and willingness should be supported by the ability to understand the relationships among the environmental, economic and social dimensions. The value of the environment has not been well recognized in the current socio-economic system. Therefore, it is necessary to deal with trade-offs among these three dimensions.

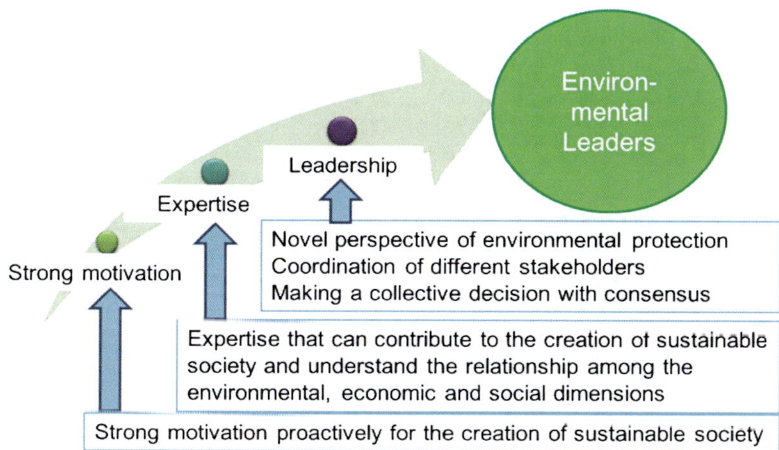

Fig. 2.8 Three key elements of environmental leaders (http://www.env.go.jp/press/press. php?serial=9516)

Expertise is also required for environmental leaders. It is useful to summarize the abilities and knowledge that are required to deal with environmental issues as an expert. The author had an experience to discuss the required knowledge and abilities of environmental engineers as a committee member for Japan Accreditation Board for Engineering Education (JABEE). Table 2.2 lists the key criteria for expertise to be acquired by university graduates. These abilities and knowledge are essential for dealing with environmental issues and to identify problems so that possible solutions can be proposed. The integration of knowledge and technologies especially is one of the most key components to reach proposals for solutions and for effective decision making.

Leadership is a special key element for leaders having expertise. Environmental leaders are expected to transfer their knowledge, information and technology for solving environmental problems. They need several leadership abilities, including the ability to:

- Capture the multi-faceted holistic environment, economic, social aspects
- Organize a novel perspective on environmental protection
- Collaborate with stakeholders by listening to their opinions and helping to address their interests
- Reach a collective decision using consensus building.

2.3.3   Education Experience During APIEL Thailand Unit 2012

In August 2012, APIEL organized a field exercise unit with the theme on "Sustainable Urban Water Management: special focus on flood management in Bangkok (Thailand Unit 2012)". During the unit, teaching staff tried to provide a

Table 2.2 JABEE criteria for expertise in the field of environmental engineering

1. Ability to understand and appropriately address the fundamental principles of environmental management, conservation, improvement, remediation and reduction of environmental load
2. Ability to observe, acknowledge and analyze phenomena relating to the environment
3. Knowledge and abilities of applied mathematics and at least two subjects from natural sciences, mainly focusing on physics, chemistry, biology and geology
4. Ability to plan and execute surveys or experiments, to accurately analyze and examine the acquired data, and to explain the results
5. Ability to identify environmental issues, to set agendas, and to propose possible solutions by integrating knowledge and technologies

holistic view of environmental issues, inter- or trans-disciplinary and cross-cultural approaches, as well as a balance between the environmental, economic and social dimensions needed to deal with complex water management issues under flood conditions. Water management is one of the most critical environmental issues, globally and locally. Tropical regions especially often have vulnerable water resources, floods, and limited access to safe water. Thus, the unit focused on the flooding in Thailand, especially in Bangkok, in 2011.

Participants looked into the flood risk not only through the "lens" of climate change but also socio-economic change. Discussion extended to water quality and quantity with respect to water security and health risk management during floods. Lectures were given to students (on flood scenarios, impact, and control) to provide them with basic knowledge and information on flood issues and their management. Technical visits were arranged to learn from different stakeholders, including Rojana Industrial Park which was damaged by the flood, Royal Irrigation Department, Bangkok Metropolitan Administration, and the World Bank. Students learned about flooding and management policies from different stakeholders, deepening their understanding through roundtable discussions. In addition, interviews with citizens covering water access and management during floods were also arranged through assistance from governmental officials.

To achieve the goals of the course, group working time was reserved for students to have in-depth discussions and to making proposals. Three working groups were organized with different themes selected by the students: flood management and land use (Group 1), access to basic human needs in case of flood (Group 2), and disease prevention after floods (Group 3). During the group work, students were expected to make proposals for dealing with their theme.

For example, Group 3 focused on the challenging issues of providing basic public health needs for disease prevention. They learned to (1) identify the problems of accessing basic human needs immediately after the 2011 Thailand flood, (2) analyze how early flood warnings and flood awareness influenced access and (3) propose possible measures to minimize problems of accessing basic human requirements in future emergencies. They conducted an in-home questionnaire survey and street intercept in Ayutthaya Province, which had been severely affected by the flooding. The questionnaire had two main parts: accessibility of targeted public health needs including early warning and flood awareness.

The Thailand Unit was successfully conducted in collaboration with the Asian Institute of Technology (AIT). Students from UT and AIT were mixed to form groups to tackle their own tasks in the field. The outcomes of the three groups were presented as posters in the 10th International Symposium on Southeast Asian Water Environment held in Vietnam in November 2012.

This type of APIEL field exercise should be designed to provide structured knowledge, as well as the practical skills and experiences that are needed for fostering young professionals with strong motivations to meet environmental challenges. Through the education experience of the Thailand Unit 2012, faculty members also learned how to better organize the course program and how to stimulate student discussions during group work. For example, faculty members in many cases, made appropriate comments and suggestions on students work according to their background and knowledge; on the other hand, faculty members observed the obstacles and struggling among students group work from time to time.

During the group work and presentation meetings, some students showed a marked improvement in integrating information and knowledge and coordinating different opinions through good communications with other students. They had acquired the high level of skills needed to explain their ideas clearly and logically while using good examples. This demonstrates to us that collaborative work with sufficient discussion time in a working group environment provides the essential opportunity to strengthen and expand the abilities that are required for environmental leaders.

2.4 Strong Leadership in a Task Force After the Tsunami[8]

2.4.1 Background

A massive tsunami hit the coast of northeastern Japan following the Great East Japan Earthquake on March 11, 2011. This was a tragic disaster for Japan, and many Japanese volunteered to help people in the devastated area. As a professional in environmental engineering, the author had serious concerns over the water environment in the area a few weeks after this tragic event. Although critical problems such as food, medical care etc. were being dealt with by various organizations including volunteer groups, water environment issues remained unsettled and drew less attention as they seemed to be less urgent.

The earthquake had a profound impact on the Japanese people, and the scientific community was no exception. *The first emergency recommendation regarding the response to the Great East Japan Earthquake* was published by the Science Council of Japan on March 25, 2011. The Japan Society on Water Environment

---

[8]This section is based on the personal experiences of one of the authors, Hiroyuki Katayama from Department of Urban Engineering, Graduate School of Engineering, The University of Tokyo, Japan.

Fig. 2.9 Wall at wastewater treatment plant collapsed by tsunami

Fig. 2.10 Sample collection at wastewater treatment plant hit by tsunami

(JSWE) also responded and formed a task team. The task team, together with the Tohoku (northeastern Japan) area branch of JSWE, proposed two fields for study groups to work on. These are called the Study Group on Health-Related Water Microbiology (HRWM) and the Study Group on Wetlands and Coastal Areas.

The author is a secretary of the HRWM study group, and was appointed as a core researcher to lead the Study group. This is a rare situation for a researcher, since this activity is not only related to research but is also closely associated with social problems.

2.4.2 Key to Success: Integration of Multi-Stakeholders

Our research team from the Department of Urban Engineering, UT had been conducting onsite field surveys in foreign countries, including collecting samples and analyzing microbial water quality, as well as simple water quality parameters. This experience was a great help for organizing field surveys in Tohoku, where no adequate experimental facility was available at the time. For the field surveys, a virus concentration method was developed and applied to a large volume of water and then modified for application outside the laboratory. Many things had to be done quickly for the field survey, and teamwork was especially important. However, students on a research team should have independent research themes, and most of them had independent scopes of research.

The culture in our research team includes the practice of "on-the-sampling training." For a water quality survey, new graduate students help with surveillance, while senior laboratory staffs instruct and advise them. Unlike lectures at the university, this advice contains unwritten knowledge about the area of the surveillance. Experience outside the laboratory also allows the students' eyes to be wide open to the real situation and gives them the opportunity to view their own research from a wider perspective. The senior members experienced this training before, and they are transferring their experience to the new students. Through day-by-day work in this laboratory culture, mutual trust was developed among the team members; the team is now always ready to receive a request of surveillance.

Communication is also an important factor. Advice from a wide range of people was gathered by senior professors from different universities in the JSWE headquarters, helping to further improve communications with stakeholders. Stakeholder groups included the local wastewater departments, and the local fisheries departments and fishermen's cooperatives in Miyagi Prefecture who all responded positively to our involvement. With strong support from these stakeholders, our proposal was submitted to the Ministry of Land, Infrastructure, Transport and Tourism (MLIT), and to the Ministry of Environment. The two ministries agreed to give us funds for research and surveillance of the wastewater system and the coastal area that receives discharge from the sewer systems.

2.4.3 Key to Success: Application of Scientific Knowledge

The first step was to select a target site for the field survey. Ishinomaki City was selected for the following three reasons: a lot of people remained producing wastewater, commercial fishing had restarted, and the wastewater treatment plant was not yet functioning. The task team set our target as affordable treatment, which can be applied in the rehabilitation stage before the full recovery of the treatment plants.

The MLIT, in charge of sewer systems, insisted on adding chlorine to the untreated wastewater to meet the coliform standard for effluent. However, this policy does not work for microbial water safety because chlorine is not effective against all pathogens, including viruses and protozoa. The task team pointed out the weak point in that policy, and tried to find alternate ways to achieve a relatively safe water environment.

Our concept was that given the urgent situation the treatment would not be perfect, but should be affordable. Usually wastewater is treated biologically over a relatively long retention time, but in this unusual case physicochemical treatment might be better able to achieve microbially safe water in a short time. Instead, the target among the water quality parameters should be narrowed down only to pathogen control, which is more urgent and important under the circumstances in a disaster area (Actually, on some occasions, nutrient or organic loading is not always harmful to the environment, but pathogens have to be removed in any case).

Our experience was useful in searching for an appropriate treatment for storm water combined with untreated wastewater, which is discharged into open bodies of water when the sewer overflows due to an overabundance of rainwater. A combination of coagulation-sedimentation and ultraviolet disinfection was tested and found to be the most effective method for producing relatively safe water.

Polysulfuric ferrite, a waste byproduct from the iron and steel industry, was selected as the coagulant. It was also used as an anti-odor agent for municipal wastewater. Polysulfuric ferrite is used for organic wastewater treatment in the food industry, among others. The proposed method successfully removed suspended solids and turbidity from the influent, and achieved a good amount of bacterial and viral removal when used together with ultraviolet light disinfection. This result was presented at a wastewater works conference.

Our method was not applied in reality in Ishinomaki City because it requires waste sludge management after the coagulation and sedimentation. The sludge management system was restored almost at the same time as the other water treatment facilities, and the wastewater treatment plants selected a biological treatment. The method the task team developed will be used in the future or in other recovery areas in the Tohoku region. The sampling campaign for water quality in the coastal area is still contributing to future water quality management plans from the viewpoint of microbial water safety. During the course of the project, much support was given to us by senior professors, stakeholders, and laboratory staff. Overall the author experienced a strong leadership applied under such a tragic disaster by integration of multi-stakeholders and application of scientific Knowledge.

References

1. Redekop BW (2010) Leadership for environmental sustainability. Routledge, New York
2. Heifetz R (1994) Leadership without easy answers. Harvard University Press, Cambridge
3. Antonakis J, Cianciolo A, Sternberg R (2004) The nature of leadership. Sage, Thousand Oaks
4. Chawla L, Cushing DF (2007) Education for strategic environmental behavior. Environ Educ Res 13(4):437–452
5. Berry KJ, Gordon JC (1993) Environmental leadership. Island Press, Washington, DC
6. Palmer JA (1998) Environmental education in the 21st century, theory, practice, progress and promise. Routledge, London
7. Gunderson L, Light S, Holling C (1995) Barriers and bridges to the renewal of regional ecosystems. Columbia University Press, New York
8. Moser S, Dilling L (2007) Creating a climate for change: communicating climate change and facilitating social change. Cambridge University Press, Cambridge
9. Taylor AC (2008) Leadership in sustainable urban water management: an investigation of the champion phenomenon within Australian water agencies. Report no. 08/01, National urban water governance program, Monash University, August 2008. ISBN: 978-0-9804298-5-5
10. Danter K, Griest D, Mullins G, Norland E (2000) Organizational change as a component of ecosystem management. Soc Nat Resourses 13:537–547
11. Burns J (1978) Leadership. Harper Perennial, New York
12. Greenleaf R (2002) Servant leadership: a journey into the nature of legitimate power and greatness. Paulist Press, New York
13. Western S (2008) Leadership: a critical text. Sage Publication, Los Angeles
14. Gordon JC, Berry KJ (2006) Environmental leadership equals essential leadership. Yale University Press, New Haven & London
15. Flannery BL, May DR (1994) Prominent factors influencing environmental activities: application of the environmental leadership model. Leadersh Q 5:201–221
16. Akiyama T, Li J, Onuki M (2012) Integral leadership education for sustainable development. J Integral Theory Pract 7:72–86
17. Wilber K (2000) A theory of everything: an integral vision for business, politics, science and spirituality. Shambhala, Boston
18. Wilber K (2007) Integral spirituality: a startling new role for religion in the modern and postmodern world. Shambhala, Boston
19. Esbjörn-Hargens S (2006) Integral education by design: how integral theory informs teaching, learning, and curriculum in a graduate program. ReVision 28:21–30
20. Inoue M, Kato Y, Moriya K (eds) (2007) Oasis chiikishi ronsou: Kokuga ryuiki 2000 nen no tenbyou. Shoukadoh, Kyoto (in Japanese)
21. Nakawo M (2011) Oasis chiiki no rekishi to kankyou: Kokuga ga kataru hito to shizen no 2000 nen. Bensei, Tokyo (in Japanese)
22. Sakai A, Inoue M, Fujita K, Narama C, Kubota J, Nakawo M, Yao T (2012) Variations in discharge from the Qilian mountains, northwest China, and its effect on the agricultural communities of the Heihe basin, over the last two millennia. Water Hist 4:177–196. doi:10.1007/s12685-012-0057-8
23. Akiyama T, Sakai A, Yamazaki Y, Wang G, Fujita K, Nakawo M, Kubota J, Konagaya Y (2007) Surfacewater-groundwater interaction in the Heihe River Basin, northwestern China. Bull Glaciol Res 24:87–94
24. Konagaya Y, Xinjirt, Nakawo M (2005) Chugoku no kankyo seisaku Seitai Imin: midori no daichi Uchi-Mongol no sabakuka wo fusegeruka? Showado, Kyoto (in Japanese)
25. Hidaka T, Nakawo M (eds) (2006) Silkroad no mizu to midori ha dokohe kietaka? Showado, Kyoto (in Japanese)
26. Nakawo M, Konagaya Y, Huhbator (eds) (2007) Chugoku henkyo chiiki no 50 nen: Kokuga ryuiki no hitobito kara mita Gendaishi. Toho-shoten, Tokyo (in Japanese)

27. Nakawo M, Qian X, Zheng Y (2009) Chugoku no mizu-kankyo mondai: kaihatsu no motarasu mizu-busoku. Bensei, Tokyo (in Japanese)
28. Wang G, Liu J, Kubota J, Chen L (2007) Effects of land-use changes on hydrological processes in the middle basin of the Heihe River, northwest China. Hydrol Process 21:1370–1382. doi:10.1002/hyp.6308

Chapter 3
APIEL Compulsory Course: Environmental Challenges and Leadership in Asia

Motoharu Onuki and Kyoungjin J. An

Abstract APIEL compulsory course, Environmental Challenges and Leadership in Asia (ECLA) discuss the human sustainability on earth by looking into the critical environmental issues in Asia and defines Environmental Leadership along with the environmental literacy and skills which are highly required in developing a sustainable society. Since many countries in Asia are densely populated and undergoing various development stages, Asia stands as the core region when considering our future global sustainability.

The feature of this course is the combination of lectures and case studies including interactive dialogues between teachers and students, group discussion and presentation in the classroom. Case studies were provided to understand discourses on industrialization (Minamata disease case), globalization (China's air pollution case) and recent urbanization (Korea's reclamation case). Students experientially learn communication, facilitation, and leadership skills by resolving tension between students from different academic fields. Therefore, this course provides platform for the field exercises where students can practice what they learn in the class.

Keywords China's environmental problems • Global environmental problems • Industrial pollution problems • Leadership education • Minamata disease • Urban environmental problems

M. Onuki
Graduate Program in Sustainability Science, Graduate School of Frontier Sciences,
The University of Tokyo, 5-1-5 Kashiwanoha, Kashiwa, Chiba 277-8563, Japan
e-mail: onuki@k.u-tokyo.ac.jp

K.J. An (✉)
Asian Program for Incubation of Environmental Leaders (APIEL),
Department of Urban Engineering, Graduate School of Engineering,
The University of Tokyo, 7-3-1 Hongo, Bunkyo-ku,
Tokyo 113-8656, Japan
e-mail: ankj@env.t.u-tokyo.ac.jp

T. Mino and K. Hanaki (eds.), *Environmental Leadership Capacity Building in Higher Education*, DOI 10.1007/978-4-431-54340-4_3,
© The Author(s) 2013

3.1 Introduction

Asia is showing increasing signs of demographic fatigue in its quest to provide for all the needs and wants of its exploding population. Of the world's more than six billion people, approximately 3.5 billion live in Asia. Two Asian countries combined—China and India—account for more than one-third of the global population; each of these nations has for more than one billion people. Environmental degradation has been one of the most visible manifestations of this evolving struggle. However, in most Asian countries, environmental activism or leadership is a fairly recent phenomenon. For example, in China, the government by the mid-1990s had started to tolerate environmental activism, as long as it focused on solving specific problems rather than criticizing government policies and decisions. Two prime indicators of environmental legislation and enforcement agencies were introduced in the ECLA class (see Fig. 3.1) to illustrate Asia's institutional level of environmental leadership and reform.

Japan was the first country in Asia to become industrialized and to face serious environmental disasters such as Minamata disease, so it was the first Asian country to pass basic environmental laws (1967) and to establish a ministry of the environment (1971) as shown in Fig. 3.1. Singapore established its own environmental ministry 1 year later (1972), with China and Malaysia following in 1974. Since then, the second wave of environment ministries in Asia occurred in the late 1980s and early 1990s. In most Asian countries, basic environmental legislation lagged behind the introduction of environment ministries.

Asia is the fastest-growing region measured by population growth as well economic development (though countries are at different stages), thus Asian economies provide many hints to our students to deal with evolving environmental

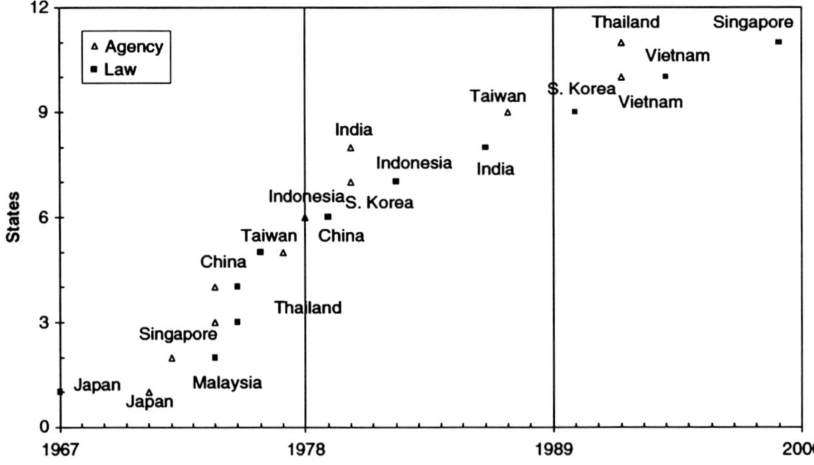

Fig. 3.1 Establishment of environmental governance, 1967–2000 (*Source*: UNEP-RRCAP, 2003)

issues associated with a wide range of development and the required leadership skills. Therefore, this course aims to review environmental issues in Asia from the history to up to date; to tackle such problems with the key of sustainably and finally to foster future environmental leaders highly required in developing a sustainable society.

The rest of this chapter consist of three sub chapter as follows: first, Sect. 3.2 describes featured ECLA's education methods in the class mainly though creating a vision, interviewing environmental leaders and debate. Second, Sect. 3.3 provides details material used for three case studies developed for ECLA. Authors carefully selected prominent case studies from discourses on industrialization in 1960s with a typical example of an industrial pollution problem that Japan has experienced (Minamata disease case); globalization (China's air pollution case) with climate change; and recent urbanization in great needs of public participation with Korea's Cheonggyecheon Restoration case in 2000s. Third and finally, we discuss and summarize main features of the compulsory course followed by overall discussion and conclusion.

3.2 Education Methods in ECLA

Course syllabus was developed and modified each year over the past 4 years in order to meet objectives of the course as shown in Fig. 3.2.

The following section describes the details about featured ECLA's education methods in the class mainly though creating a vision, interviewing environmental leaders and debate.

3.2.1 Creating a Vision Through Group Discussion

During the very first ECLA class, the authors asked students to choose one environmental project from his or her own country and write an action plan for a funding proposal to the World Bank. About 40 students met and formed 7 groups: 5 from Department of Urban Engineering (UE) and 2 from Graduate Program in Sustainability Science (GPSS), The University of Tokyo. They then negotiated to decide on only one project. In doing this, the students were required to identify the real-world problems and create a vision through collaboration with various stakeholders in each locality. Proposals from UE and GPSS students are summarized in Table 3.1 and are categorized into ten emergent environmental issues in Asia taught by the author based on a review of the literature and authors' experiences.

As we can see from the table, students address the issue of population growth associated with poverty and land use, agriculture, forest, water, and energy. However, biodiversity, air quality, global warming, and environmental activity have not been tackled. The teachers another concern is that in the midst of competing demands,

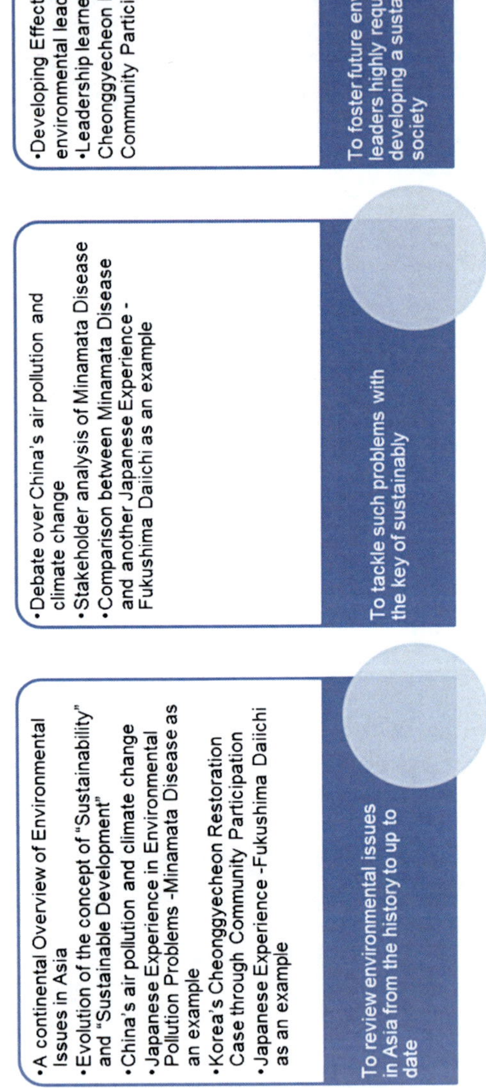

Fig. 3.2 Course syllabus

Table 3.1 Emergent environmental issues in Asia and Students' interest in the environment (ECLA class of 2012)

Author (from Hillstrom and Hillstrom [1])	UE students	GPSS students
Population and land use Food security, consumerism, urbanization and migration, and land management, etc.	Poverty issue: education, training skills in India Sustainable regional rehabilitation in the Tohoku region in Japan (land use)	
Biodiversity Land conversion and fragmentation Alien species by trade, in potted plants and shipment of crops Overharvesting and the trade in Asian wildlife *Park, preserves, and protected areas* Conservation of biodiversity National parks, wildlife sanctuaries, and nature reserves		
Forests Sustainability of forest, agriculture and other systems based on ecosystem dynamics People became more knowledgeable about the limits of the earth because of natural systems Sustainability became a very old idea and social goal		Forest conservation: Indonesia Activity: eco-tourism
Agriculture Continual increase in the world's demand for food Attempts to slow human population growth in order to reduce future food demand Intensification of agricultural production Increased use of pesticides and herbicides, decreased soil quality, conversion of forest and wetlands to agriculture, damage to watersheds from runoff, and the deterioration of rural communities	Promotion of sustainable agricultural use and management by education and training in Vietnam (less fertilizer)	Promotion of organic farming in China: solve problems of nitrate groundwater contamination
Freshwater Rehabilitation of major rivers ruined by decades of heavy pollution Acutely vulnerable to weather-related disasters Industrialized countries (e.g., Japan, Singapore, South Korea and Taiwan) clearly have fewer problems providing access to safe water, while in developing country access to safe drinking water is often a luxury, sometimes only available to the affluent	Water issue: implementation of end-of-pipe technology for industry wastewater treatment in the Philippines	

(continued)

Table 3.1 (continued)

Author (from Hillstrom and Hillstrom [1])	UE students	GPSS students
Ocean and coastal areas		
Population surges bring marine pollution		
Japan is another nation that has launched vigorous ocean protection programs, yet during the 1960s and early 1970s the country's coastal waters were extremely fouled with toxins and other waste, such as the tragic situation at Minamata Bay		
Energy and transportation	Sustainable resource management in Japan	
Limited energy source from coal, oil, and natural gas		
Alternative energy, such as hydro and nuclear power		
Renewable energy		
Transportation		
Air quality and the atmosphere		
Fossil fuel consumption in the energy, transportation, and industry sectors		
Environmentally destructive mining practices, and slash-and-burn deforestation		
Doubling of per capita commercial energy use in most regions of Asia between 1975 and 1995, a period when industrialized countries were making marked strides in improving their energy use		
Environmental activism		
In most Asian nations, environmental activism is a fairly recent phenomenon		
Global environmental citizenship in the 1990s, particularly in the period leading up to the Kyoto Climate Change Conference (1997)		

the priorities are for easy or low cost tasks. In other words, we have limited leadership capacity to tackle these pressing issues. Leaders need to provide a vision and solutions at this stage, though many of the problems seem formidable. However, this exercise allowed future leaders (the students) to practice setting clear priorities from among the competing needs and demands.

3.2.2 Interview Environmental Leaders

Gordon and Berry [2] stressed the difference between environmental leadership from other types of leadership. The Central premise is that the difference of the environmental leadership is based on the unique characteristics environmental problems that require long time to solution, complexity and uncertainty, emotionally charged situations, an incomplete and scattered science base, and the necessity for integration.

This section discusses the relationship between leadership and environment issues to be solved by conducting survey and interview students in class. The authors reviewed successful examples of environmental leaders in Asia with students and used survey questions for young leader's perspectives on environmental leader. The survey questions are used to discuss how students see today's environmental leaders: included important social trends or conditions affecting today's environmental leaders, major barriers facing today's leaders, how leaders emerge in their individual organizations, and the five characteristic the students think are most important for today's leader. Based on this subjective analysis, more objective statements about leadership were written. The results are summarized in Table 3.2.

Most of the respondents agreed with statements 2 and 4: "Men and women often have different environmental leadership skills and styles" and "Different people lead in the organization at different times." These responses showed that future environmental leaders clearly realize the complexity of environmental issues, and that what are required of them are flexibility and the ability to work with different people at different times. In other words, for the development of a sustainable future, education or training in an organization should be multifaceted. Gordon and Berry also explained that leaders perceive leadership as a function diffused throughout an organization, not the sole property of a leader based on his/her position or title, but rather every member of an organization must be prepared to lead when his/her turn or time comes. More interestingly, the respondents agreed more positively with statement 7 "Leaders today are more process, rather than product, oriented" and also with statement 12 "Leadership today is more difficult than in the past." This is in line with the reason why APIEL puts a high value on communication among the stakeholders to reach a consensus and promote a clear vision: APIEL emphasizes the development of "process and participation" leaders rather than "command and control" leaders.

The most controversial survey results were those for statement 9 "Command and control leadership is still necessary at times for environmental leadership" and statement 10 "Environmental leaders cannot lead without authority." Half of the respondents seem to agree with these statements. This suggests that participatory, collective leadership in this context is not yet mature and requires time and energy. The respondents acknowledged that control leadership is still required to drive change.

Finally, the results for statement 11 verified our strong belief that leadership skills and styles can be learned, and among these skills, the communication skill is an important key factor for leadership development. The following section describes how in-class teaching helps to develop communication and debating skills, using a case study: China's challenge for climate change.

Table 3.2 Student response to survey on environmental leadership (ECLA class of 2011, 33 respondents)

	Strongly agree	Agree	% Agree	Neither	Disagree	Strongly disagree	% Disagree	Total
1. In most organizations, environmental leadership needs to be attributed to one identifiable person	3	12	*45.5*	6	9	3	*36.4*	33
2. Men and women often have different environmental leadership skills and styles	3	18	*63.6*	6	6	0	*18.2*	33
3. The problem with today's environmental leadership is that no one is accountable	4	8	*37.5*	11	7	2	*28.1*	32
4. Within an organization, different people will lead at different times	8	16	*72.7*	8	1	0	*3.0*	33
5. Important environmental leadership skills usually stay constant regardless of the situation	2	13	*45.5*	4	10	4	*42.4*	33
6. In general, we have less evident environmental leadership today than in the past	0	3	*9.1*	8	19	3	*66.7*	33
7. Leaders today are more process, rather than product, oriented	2	15	*51.5*	11	4	1	*15.2*	33
8. Environmental leadership is basically different than other kinds of leadership (e.g., leadership in business or the military)	10	13	*69.7*	1	9	0	*27.3*	33
9. "Command and control" leadership is still necessary at times for environmental leadership	5	13	*54.5*	12	4	1	*15.2*	33
10. Environmental leaders cannot lead without authority	6	11	*53.1*	6	8	1	*28.1*	32
11. Environmental leadership skills can be learned	8	23	*93.9*	2	0	0	*0.0*	33
12. Environmental leadership today is more difficult than in the past	13	11	*72.7*	6	3	0	*9.1*	33

Percentage value were indicated in italic

3.2.3 Communication Practice Trough Debates

3.2.3.1 Global Communication

Most APIEL students agree with the idea that communication, listening, and interpersonal skills are the most important leadership characteristics. Global communication for climate change made a breakthrough following the Kyoto Protocol to the United Nations Framework Convention on Climate Change: countries were urged to take a proactive stance and mitigate their CO_2 production according to their economic growth. The first Earth Summit in Rio in 1992 focused on biodiversity, global warming, and forest initiatives. In 2002, the second Global Summit on Sustainable Development (Rio + 20) set out to resolve the conflict created by the needs of rich and poor countries and their impacts on the global economy [3]. The environment is no longer an isolated issue; it is at the heart of our global future. By 1995, countries realized that emission reduction provisions from the Rio convention were inadequate. Two years later, the Kyoto Protocol legally bound developed countries to emission reduction targets. The Kyoto Protocol's first commitment period started in 2008 and ends in 2012. One of the greatest global concerns is China's growing contribution to climate change. (China is the second largest contributor to climate change.) China has signed the Kyoto Protocol and has initiated some projects—with EU members—under the protocol. After 4 years of negotiations, the question of what happens beyond 2020 was also answered at Durban. The decision on the green climate fund extension of the Kyoto Protocol comes into effect by 2020. These include reduction pledges for 2020 and the goal of keeping global warming below 2 °C. There will be a joint implementation by China, India, and the US.

Learning from this global communication, the difficulty becomes apparent when countries, who would all benefit from addressing climate change, accept a global legal framework. However, China's goal to generate 10% of their power from renewable sources by 2010 was unlikely to have a significant impact on its contribution to climate change.

3.2.3.2 Communication Practice in Class

ECLA students—from UE and GPSS—communicated using debates on whether China has to play a leadership role on climate change. There are some special terms used in the debates. The opinion in a debate is called a resolution. The debate team that agrees with the resolution is called the affirmative team. The team that disagrees with the resolution is the negative team. The people in charge of evaluating the debate are called judges. The author has been a moderator at these debates. Interestingly, after discussion among the students, the ECLA class was divided into negative (UE) and affirmative (GPSS) teams. In the debates, the goal has been to persuade the judge and the audience that their opinion is the most compelling. To persuade the judge and the audience, the speaker should have strong arguments, which means tough reasoning, factual support, and refutations as well as a strong manner, which includes gestures, eye contact, posture, voice, and the use of humor.

Throughout the debating exercise, scientific evidence supporting their reasoning should be clear and easy to understand. In the rebuttal, each team explains where their points stand, and where the points of the other team fall, from the social and economic perspectives. Overall, we try not to lose our senses of humor. In addition, the students are able to practice listening to understand what each speaker is saying, as well as good teamwork, and expressing their own points while refuting the other team.

3.3 Materials Used for Case Studies in ECLA

This sub chapter introduces material used for three case studies in detail. As illustrated in Fig. 3.3, based on provided material, students analyze the fact and interacted with teacher to build leadership capacity through and through.

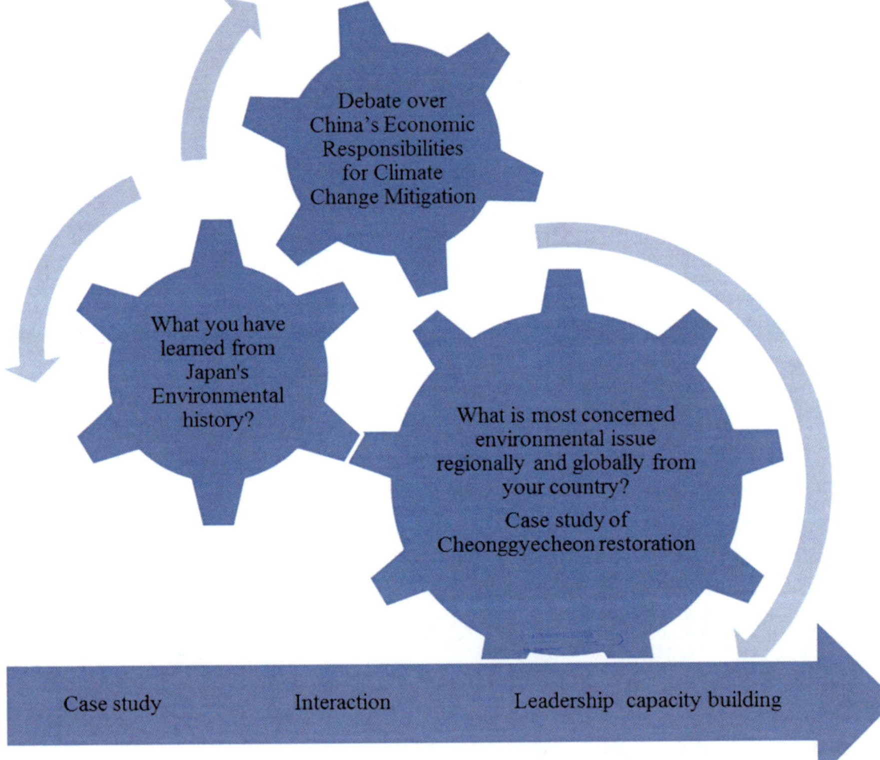

Fig. 3.3 Material used for case studies in ECLA

3.3.1 Minamata Disease and Japan's Experience with Industrial Pollution

For Japan, industrial pollution originates as environmental problems. Why they happened and how they were overcome give us several lessons for preventing new pollution problems from occurring in Japan or in other parts of the world.

Minamata disease is a typical example of an industrial pollution problem that Japan has experienced. Minamata disease, a disease of the central nervous system, is caused by eating fish and shellfish contaminated with methyl mercury compounds. These compounds were discharged into the sea at the Shin Nihon Chisso (hereafter Chisso) plant in Minamata. The compounds accumulated in the marine life [4]. Since it was one of the first severe industrial or environmental pollution problems in Japanese history, limited knowledge and lessons were available at that time. It took a long time to understand and overcome what was occurring. Minamata disease is one of the largest and longest-lasting pollution problems in Japan, and the Minamata disease issue represents nearly all aspects of pollution problems. For these reasons, this Mianamta case provides an essential case study for fostering environmental leaders.

The objective of this case study is to learn the following aspects of environmental pollution: (1) environmental pollution ultimately harms people, and protecting the environment is critical, (2) economic growth and human health/rights, (3) discrimination, the social divide, and environmental justice (4) science, technology, and society, (5) environmental science and environmental engineering, (6) environmental economics, and (7) post-modern and risk issues. This list shows that the issues surrounding Minamata disease are ongoing. It also touches on the most advanced issues related to the environment and sustainability.

3.3.1.1 Environmental Pollution Ultimately Harms People and Protecting the Environment Is Critical

The first step of this case study is to understand the tragic consequences of Minamata disease. Through this step, students learn how a polluted environment ultimately harms people and that protecting the environment is essential. The tragic experiences of the victims include not only the disease itself but also how other people treated them.

In Minamata today, several of the victims are registered as storytellers and opportunities to listen to their stories are organized. Many elementary or junior high schools are taking advantage of this storytelling for their social study courses. Storytelling is also used during training courses at universities and companies. ECLA also shows a video of the storytellers and has students read books, such as *Kugai Jodo* (Paradise in the Sea of Sorrow) written by Ishimure [5] to help them understand the victims' experiences.

3.3.1.2 Economic Growth and Human Health/Rights: Chisso's Irresponsibility

One of the reasons why Minamata disease was so widely spread was Chisso's irresponsibility. Chisso conducted an experiment called the "cat number 400 experiment" to reveal the cause of the disease by themselves. In this experiment, a cat that had been given food to which factory effluent had been added showed clear symptoms of Minamata disease. This result demonstrated that Chisso's discharge was the causative substance in the environment. It indisputably showed Chisso's responsibility for the disease, even though the pollutant had not yet been identified as methyl mercury. However, the results of the experiment were not published. Chisso placed the highest priorities on earning a profit. In order to continue operating the plant, Chisso buried the results of the experiment. Moreover, when Kumamoto University pointed out that methyl mercury was causing the disease, Chisso raised questions. It is obvious that Chisso was putting profits before people and ignoring human rights. The most important point here is to let students think about what they might have done if they had been a researcher or engineer inside Chisso.

3.3.1.3 Economic Growth and Human Health/Rights: Citizens' Awareness

We also have to remember that many Minamata citizens supported Chisso, because Minamata was heavily dependent on Chisso for tax revenue. Without Chisso, Minamata would have been a small local village in western Japan. Some citizens considered Minamata disease as someone else's problem or even as an annoyance. Others thought that a certain amount of "pollution" must be acceptable in exchange for prosperity in Minamata. Moreover, to people living in Tokyo, Minamata disease was an issue occurring far away from them and was not their concern. Many storytellers mention that Minamata disease is a disease of the modern age where people want too much convenience and too much prosperity based on industry and technology.

3.3.1.4 Discrimination, the Social Divide, and Environmental Justice

Most of the patients identified in the early stages were fishermen and their families. Since fishermen in Minamata originally came from the Amakusa Islands (on the other side of the Shiranui Sea), they were considered outsiders and were lower down in the social hierarchy. This is another reason why the patients faced discrimination. Harada [6] and Ui [7] have denied that industrial pollution created "abandoned" victims. Instead, they argue that industrial pollution problems occur where there are weak people and industrial pollution hits them. Note: today, Minamata

City has been trying to re-develop community relations under the concept of
moyainaoshi or re-connecting [4].

3.3.1.5 Science, Technology, and Society

Chisso's irresponsible attitude has been discussed in the previous section. If the
result of the cat number 400 experiments had been published in 1959, countermea-
sures for Minamata disease would have been taken and the number of victims would
have been completely different. However, another factor, from the perspective of
"science, technology and society", has also been pointed out.

When Kumamoto University was concluding that methyl mercury as a causative
substance, Chisso raised counterarguments:

1. Why was it not until 1954 that Minamata disease suddenly occurred, although
   Chisso has been operating since 1932?
2. Why did Minamata disease occur only in Minamata although many acetaldehyde
   factories had been using the same chemical process?

Once counterarguments are raised, scientists usually try to respond. In spite of a
government order forcing Chisso to submit wastewater samples, scientist tried to
answer these questions one by one, although what was urgently needed was only to
prove that factory effluent was causing the disease.

In the field of science, there are many unsolved problems, and scientists are
always doing research. However, government and citizens expect scientists to
come up with a 100% complete answer. If even 1% remains unsolved, governments often wait. They use the fact that arguments are still continuing as a reason
why they aren't taking any measures for the relief of victims. This relationship
between the scientific community and government has been called the "resonance
of scientists and government" by Sugiyama [8], and Harada [6] mentioned that the
unsolved 1% should not be used as an excuse for corporations or governments to
avoid accepting responsibility. Ui mentioned that there is no neutral standpoint
with industrial pollution [7].

3.3.1.6 Environmental Science and Environmental Engineering

The "resonance of scientists and government" was pointed out in the previous
section. Relief for victims should start even if some issues remain. At the time, the
cat number 400 experiments were enough evidence to start relief measures.
However, scientific study is still necessary, especially for preventing the same mis-
takes from occurring in the future; scientific study must continue.

The two questions raised by Chisso were finally answered by Nishimura and
Okamoto [9] more than 40 years later. They revealed the chemical reaction path-
ways, determined the kinetic parameters of methyl mercury production under several conditions, and finally determined that switching the oxidant from permanganate

to nitric acid in 1951 was the cause of the sudden increase of methyl mercury discharge. This was the answer to question 1 (above). However, using nitric acid as an oxidant was the normal practice in acetaldehyde production, which didn't cause any problem in other factories or companies in Japan. Nishimura finally found that the high concentration of chlorine ions at the Chisso factory was the key difference. When chlorine ion concentrations are low, during the normal process (using nitric acid as an oxidant) only tiny amounts of methyl mercury are produced, but when the chlorine ion concentration is high, methyl mercury is vaporized and easily discharged to the outside. That was the answer to question 2 [9]. Even though it took 40 years to answers these questions, they are still important examples of environmental scientific research, and worth studying.

3.3.1.7 Environmental Economics

Once the environment is polluted, it takes a long time and a huge cost for recovery. When pollution diseases and victims appear, it is nearly impossible to recover from the damage, and compensation can be enormous. From the viewpoint of environmental economics, the cost for preventive measures, including wastewater treatment and developing production processes that don't discharge wastewater, is cheaper than the cost of compensation and recovery of the environment [10]. The quality of the environment started to dramatically improve once this fact was widely recognized by industry, and after environmental regulations were established. This is the Japanese government's message to other countries, especially those who are currently developing rapidly.

It is important to remember that for establishing the legal framework for compensation and for developing environmental regulations we had to see victims suffering from the disease and being discriminated against, as well as having to experience lawsuits. How to reach the point of realizing that prevention is cheaper than compensation without these bitter experiences is the most important point in the class.

3.3.1.8 Post-Modern Issues: Health Risks and Low-Level Exposure

As discussed above, most Japanese people consider the Minamata disease issue solved and finished, and that it is full of lessons to be learned. However, the Minamata disease issue has not yet been finalized. Since 1968, when the Japanese government officially declared that Minamata disease was an industrial pollution problem caused by Chisso and Chisso agreed to compensation payments, many patients have been applying to be certified as victims so that they can be paid. However, many of these claims were rejected or were left pending. This is because the criteria for Minamata disease certification have been based on exhibiting the symptoms of Hunter-Russell syndrome. Hunter-Russell syndrome was discovered after methyl mercury poisoning was found in British factory workers who were producing disinfectants for seeds. It is characterized by five symptoms: numbness and pain in the extremities, dysarthria, ataxia, auditory disorders, and concentric constriction of the visual field. However, Minamata disease was caused by methyl mercury poisoning following the consumption of fish and shellfish contaminated by environmental pollution. The levels of contamination and pollution varied; therefore the way in which each symptom appeared also varied. This is why the criteria for official certification are considered to be too strict. Sufferers who did not have acute symptoms—perhaps only one or two symptoms (sensory disorder, numbness, etc.)—were left uncertified as victims.

Since several thousand applications were rejected or left pending in the 1990s, the Japanese government decided in 1995 to relieve these "uncertified" patients by paying their medical expenses and lump-sum amounts. Patient support groups are requesting large-scale medical checkups and revised criteria for official certification of Minamata disease. However, the Japanese government decided to review uncertified patients using different criteria instead of changing the official legal criteria. For political, social, and economic reasons, the criteria for diagnosing Minamata disease are not well defined and detailed scientific knowledge on the minimum level of Minamata disease (or methyl mercury poisoning) is still not available. More scientific data and large-scale medical checkups are necessary to understand the overall picture of Minamata disease as well as the entire mechanism for environmental pollution caused by the methyl mercury discharged by Chisso. Why hasn't this been done? Of course, the extent of suffering of "uncertified" Minamata disease patients is generally smaller than typical acute Minamata disease patients found in 1950s and 1960s. Nonetheless, when we try to definitively answer this question, we may well face renewed discussions similar to those outlined above. This is why the authors think the Minamata disease issue has not yet been put to rest and is still an example of one of the hottest topics in environmental issues: health risks and low-level exposure.

3.3.2   China's Air Pollution Case Over Last Three Decades

3.3.2.1   China's Economic Growth

Over the last three decades, why has China's environment been drawing so much worldwide attention, along with its rapid economic growth? The rise of China as an economic power is one of the most remarkable stories of the latter half of the twentieth century. There were annual growth rates of from 8% to 12% of GDP, and by the end of 2011 China had become the second largest economy, after the United States. However, the progression was different from other developed countries. For instance, Japan progressed through the stages of industrialization, urbanization, and globalization over a much different time span. But China's economy boomed in the face of a raging wind of concurrent industrialization, urbanization and globalization. In China, industrialization depends heavily on intensive energy and resource use, urbanization is not based on a sustainable model, and globalization, through exporting many goods and absorbing manufacturing from other countries, has led people to derisively call China the "world's factory." At the same time, this growth has occurred without much consideration for the environment. Growth has placed tremendous pressure on China's environment. Building on centuries of environmental degradation and pollution, the very rapid industrialization of the last quarter century has contributed to some of the highest rates of air and water pollution in the world, as well as severe land degradation, and a range of other emerging resource challenges; among them all, the most visible of China's environmental challenges is air pollution [11].

3.3.2.2 China's Air Pollution

China's air pollution is due to an overwhelming reliance on coal for energy production. China relies on coal for approximately 70% of its energy, consuming about 1.96 billion tons in 2004. The carbon dioxide emission rate reached 1.6 GtC (gigatons of carbon or 10^{15} g carbon) per year in 2006 [12], and China has become the world's largest emitter of sulfur dioxide. Moreover, by 2020, transportation experts anticipate that China will have at least 110–160 million cars on its roads. Today, China is attempting to push forward with renewable energy and with alternative fuels, such as compressed natural gas.

3.3.3 Korea's Cheonggyecheon Restoration Case Through Community Participation

On July 17, 2009, 5 years after the completion of the Cheonggyecheon project, the *New York Times* reported that the restoration was part of an expanding environmental effort in cities around the world to "daylight" rivers and streams by "peeling back pavement" that was built to bolster commerce and serve automobile traffic decades earlier [13].

3.3.3.1 History of Cheonggyecheon and Leader's Attitude

In Korean, Cheonggyecheon means "clean water stream" and this area of Seoul once had deep significance for the people living there. It runs through the city center. Seoul was chosen as the capital of Korea more than 600 years ago, and was positioned among four mountains: ones to the east, west, north, and south. Cheonggyecheon provided a cultural space and shelter before it was covered over with a road. For a long time, Cheonggyecheon was both a flood control channel and a place to wash laundry. However, by the 1940s, the stream had begun to fill up with sewage and trash, and it became a slum area after the Korean War. Since 1958, the stream was gradually covered over with a concrete road for two main reasons: filthy water (and an unsafe environment that made people lose interest in the area)

Table 3.3 History of Cheonggyecheon

	Stream status	Leaders' attitude to waterway control
1400s–1500s	Natural and artificial stream	People's interest
1600s	Stream sustaining the city life; abandoned to natural control	Scholar's interest
1700s	Dredging, controlled stream	Strong leadership from king
1800s	Less controlled stream	
1960s	Unsanitary sewer and slum	Less interest
1970s–1980s	Highway for industrialization	Development oriented
1900s–2000s	Old market selling everything; declining traffic, population and environmental contamination	

and pressure from rapid economic development in the 1960s that required more roads and an easing of traffic congestion. That is why, in 1968, the Cheonggye Elevated Highway was constructed over Cheonggyecheon. Table 3.3 summarizes the history of Cheonggyecheon from two aspects: streams status and leaders' attitude to its control.

3.3.3.2 Paradigm Change for Urban Development

There has been a paradigm change and a new vision for this area of Seoul. In 2003, after about 40 years of neglecting Cheonggyecheon, the restoration project was begun following the introduction of environmentally sound and sustainable development (ESSD) with the key idea of urban development sustainability. The paradigm of urban planning is shifting from a growth-oriented or quantitative growth model—by expanding urban spaces based on transport requirements, large-scale investment, development effect, facilities development and strategic point development—to a growth management or qualitative growth model based on humanism and naturalism, pedestrian requirements, quality of life improvement, benefit/management, developing "soft" programs, and a network system.

The project was led by the City of Seoul, especially by the mayor, Lee Myung-bak (the President of Republic of Korea later), who showed a clear vision. More than 70% of the citizens supported the restoration. The vision was to create an environment-friendly city space with an emphasis on nature and people, solving the safety problem related to covered structures, recovering the 600-year-old capital city's historic and cultural significance, promoting balanced regional development, and providing a "hands-on eco experience" to millions of citizen [14].

3.3.3.3 Implementation of Cheonggyecheon and Public Participation

This restoration project had wide-ranging public participation. The master plan included dismantling the elevated highway and the structures covering the stream, moving some structures to other places, and building the infrastructure to help restore the stream, including a sewage system, roads, bridges, landscaping, and lighting [15]. During the development process, the general public, specialists, and interest groups formed a citizens' committee and offered their opinions. On the other hand, those opposed criticized the government-led restoration project, and coalesced around main three groups [16]: (1) The first to step forward were academic groups such as the Environmental Sociology Association, the Korea Planners Association and the Environmental Impact Assessment Society. They organized a series of academic debates that disclosed the theoretical and practical drawbacks associated with the civil engineering-focused restoration, such as insufficient ecological considerations, undemocratic procedures, excessive commercial re-development of the surroundings and an unsustainable water flow. (2) There was a strong public and NGO coalition movement, organized by a citizens' coalition, against what was regarded as an undemocratic, anti-environmental and politically manipulated restoration. (3) The last opposing force was made up of more than 60,000 merchants along Cheonggye Road—employing 800,000 workers—plus a large number of illegal street vendors. Most shopkeepers were worried about the loss of business during the construction period. The merchants were vehemently opposed to the restoration plan, so they joined the street vendors in sometime violent demonstrations that threatened the project. The merchant group was not only the most recalcitrant but also the most important partner from whom Mayor Lee needed to gain consent in order to carry out the project as planned. Unlike other opposition groups, however, the merchant group's major concern was to safeguard and maximize their private interests, but the group was too fragmented to be effective.

It is most desirable for a stream to receive water from its upper reaches. However, the arguable fact still is that Cheonggyecheon requires additional flow from the Han River to maintain a certain depth throughout the year. This is likely to continue until advanced technology is in place to treat environmental hormones, smell, and the foaming of wastewater, as well as a change in people's perceptions for reusing wastewater.

3.3.3.4 Leadership

Leadership and some measures were necessary to persuade the opposition groups. The City of Seoul took some measures to ease traffic congestion. To gauge the opinions of merchants on the impact of the restoration project on their business, the City of Seoul held public hearings and gave presentations for each commercial block. These steps helped to win over the Cheonggyecheon Residents and Merchants Council and the Cheonggyecheon Merchants Association. Seoul also conducted more than 4,000 interviews with merchants before the start of the demolition work. Based on the opinions collected, measures were devised to address complaints related to inconvenience to businesses. Additional measures were devised for stimulating business activity that considered the unique character of the Cheonggyecheon commercial district, which is made up of several business quarters with widely varying interests. In response to academics and NGOs, the Cheonggyecheon Restoration Project helped to restore the long-forgotten history and culture of Seoul. Traditional cultural activities were revived, such as "bridge stepping" on Supyogyo Bridge (when people go out onto the bridge to look at the full moon) and a lantern festival. In addition, the local bridges were restored, and fundamental safety problems were resolved for the Cheonggye Elevated Highway and some of the structures covering the stream.

3.3.3.5 After Cheonggyecheon Restoration

The impact and economic cost/benefit were also assessed. There was the most controversy over the 24-h runoff of 120,000 tons of water from the Han River. This required pumping and large amounts of energy that contribute to environmental problems, such as the greenhouse effect. The most significant change, however, was that millions of visitors went to the restored district. Cultural events have been held almost every day on the Cheonggye Plaza, located at the starting point of the waterway. The plaza has now become the most attractive cultural civic space in central Seoul and visitors are highly satisfied with their experience, as the author stated in the *New York Times* article.

The Cheonggyecheon restoration is known as the most successful recent example of public policy in Korea. Yet, at the same time, it is still at the center of a controversy because of the evaluation of its character [15]. Cheonggyecheon's restoration has helped to improve downtown Seoul and it is possible to view the project as part of the "ecological modernization" of Seoul [16]. This project was led by the strong political ambition of its initiator. According to the Seoul Development Institute, a prominent partner in this project, willingness to pay 103,309 won per household per year (8,609 won/month) or 356.2 billion won/year in Seoul was measured after this successful project. This implies that Cheonggyecheon's restoration inspired different thoughts about the quality of life and the willingness to sacrifice for the public good. Citizens are now willing to pay for non-economic value and public benefit. Thus, the social cost: construction + congestion (loss by traffic load) + waterway maintenance over social benefit: savings of repair costs (no highway maintenance) + environmental benefit (willingness to pay) was measured as Benefit over Cost (B/C) = 1.85 [17].

3.4 Features of the Compulsory Course

3.4.1 Diversity

One of unique features of the ECLA class is the student's interdisciplinary from either engineering/technology (the Department of Urban Engineering, UE) or sustainability science (the Graduate Program of Sustainability Science, GPSS), as well cultural diversity from all over the world (though the majority of the students are Asian). Students struggle with such diversity during group work, but often mutual benefit is greater and led to desirable outcome.

3.4.2 Interactive Dialogues Between Teachers and Students

Considering the speed of Asia's economic and population growth and the diversity of students, education methods in ECLA was modified from traditional type of information delivery to transmit needed information throughout the process of finding and understanding. Developing effective skills and style for environmental leaders drew the image of environmental leaders through an interview and survey to metaphor the image of leaders, and exercised the creating a vision through consensus building and setting the priorities among vision, in the class. Inclusion, or listening using all available skills and ideas were emphasized in the classroom as well. For example, debate over China's air pollution and climate change; stakeholder analysis of Minamata disease; comparison between Minamata disease and another Japanese experience of Fukushima Daiichi (Nuclear Power plant) aids students to develop the leadership capacity.

Education methods in ECLA showed how we explored the possibilities and requirements for addressing trends in environmental leadership in market-oriented, industrialized, and industrializing economies in Asia during the past several decades, and practiced essential leadership skills prior to APIEL's field exercise.

3.4.3 Case Studies

The author communicated with students to discuss the evidence for improvement (or deterioration) in environmental quality by case studies, and the lessons learned from them. Materials used for three case studies were chosen : (1) Minamata disease, a typical example of an industrial pollution problem that Japan has experienced, deal with industrialization in 1960s in Japan and many of Asia country in recent years; (2) China's air pollution case over last three decades, due to an overwhelming reliance on coal for energy production and its effect to climate change associated with current globalization; and (3) Korea's Cheonggyecheon Restoration case to understand recent urbanization in 1990s, and address the importance of public participation. Overall, communication between the authors (teachers) and students "about," "in," and "for" the environment based on case studies enables us to see how students can increase their capacity to play the role of future leaders.

3.5 Discussion and Conclusion

This series of ECLA lectures provides environmental issues and leadership lessons learned from history, as well as discourses on industrialization, recent urbanization, and globalization. As such, the Minamata disease issue can be viewed from many perspectives. Stakeholders, including professionals (scientists, policymakers, as well as government and company officials) and citizens each have their roles and responsibilities. It is important for environmental leaders in many stakeholder groups to understand these other points of view. In addition, the lessons learned are not only the ones from the past, but also from some of the ongoing problems that are directly linked with modern life.

The authors reviewed the issues to be addressed to educate students to understand and to build leadership skills, as well as the attributes and ethics associated with the complex challenges that we are facing as we prepare for a sustainable future. The lectures include interactive dialogues between teachers and students, and students among themselves, to criticize, understand, persuade, and learn throughout the journey. The authors put emphasize such practices prior to APIEL's field exercise. Because arriving at consensuses vision is always time consuming and a first barrier students have to go through and division of role at field work is of importance.

Therefore, these classes are the "seed" for the field exercises where students can practice what they learn in the class. Reflection of this education method in ECLA on each field excises will be introduced in the part two. The vivid stories of the field exercises will be outlined in Chaps. 4–7 in the second part of this book.

References

1. Hillstrom K, Hillstrom LC (2003) Asia: a continental overview of environmental issues. The World's Environments Series. ABC/CLIO, Santa Barbara
2. Gordon JC, Berry JK (2006) Environmental leadership equals essential leadership. Yale University, New Haven & London
3. Sonnenfeld DA, Mol APJ (2006) Environmental reform in Asia: comparisons, challenges, next steps. J Environ Dev Volume 15(2):112–137
4. Minamata City (2007) Minamata disease. Its history and lessons. Minamata City Planning Division, Kumamoto
5. Ishimure M (1990) Paradise in the sea of sorrow: our Minamata disease. Livial Monnet, trans. Yamaguchi Publishing House, Kyoto (First published in Japanese as Kugai Jodo: Waga Minamata-byo by Kodansha, Tokyo 1972)
6. Harada M (2004) Minamata disease Tshushima S and George ST. trans. Kumamoto Nichinichi Shimbun Culture and Information Center, Kumamoto (First published in Japanese as Minamata-byo by Iwanami Shoten Publishers, Tokyo 1972)
7. Ui J (1992) Minamata disease. In: Ui J (ed) Industrial pollution in Japan. United Nations University Press, Tokyo
8. Sugiyama S (2005) Minamatabyou jirei ni okeru gyousei to kagakusya to media no sougo sayou. In: Fujigaki Y (ed) Case analysis and theoretical concepts for science and technology studies. University of Tokyo Press, Tokyo (in Japanese)
9. Nishimura H, Okamoto T (2001) Science of Minamata disease (Minamatabyo no kagaku). Nihon Hyoron Sha, Tokyo (in Japanese)
10. Pollution in Japan—Our Tragic Experiences, Study Group for Global Environment and Economics, Ministry of Environment (2006) Annual report on the environment in Japan 2006

11. Economy EC (2006) China's Environmental Challenge Asia Studies Council on Foreign Relations, Testimony Before the U.S.-China Economic and Security Review Commission Hearing on Major Challenges Facing the Chinese Leadership
12. Zeng N, Ding Y, Pan J, Wang H, Gregg J (2008) Climate change—the Chinese challenge. SCIENCE 319:730–731
13. New York Times (2009) Peeling back pavement to expose Watery Havens, 17 July 2009
14. Shin JH, Lee IK (2006) Cheonggyecheon restoration in Seoul, Korea. Proc Inst Civil Eng-Civil Eng 159(4):162–170
15. Cho MR (2010) The politics of urban nature restoration—the case of Cheonggyecheon restoration in Seoul. Korea IDPR 32(2):2010. doi:10.3828/idpr.2010.05
16. Cho MR (2006) Developmental politics and green progress. Environment and Life, Seoul (in Korean)
17. Seoul Development Institute (2005) A study on the basic plan of Cheonggyecheon place marketing strategy. Seoul Development Institute, Seoul

Chapter 4
Leadership Development for Sustainable Urban Environmental Management: Cases in Thailand

Tomomi Hoshiko and Tomohiro Akiyama

Abstract This chapter explores future challenges to improve the design and implementation of the Thailand Unit by examining two cases of the unit conducted in 2009 and 2011. The unit covers two important issues of urban environmental management: Solid waste management and urban water use and management. Development, implementation, results and review processes of the unit are shown by comparing both cases. To examine educational effects of the unit in terms of its academic contents, group work results are shown. To improve the unit design and implementation, questionnaire survey results for the involved faculty members are shown, where the strengths and weakness are clearly indicated. To review the effects of the unit participation on leadership development in individual students, their feedback comments are shown and serve as proof of the unit's achievements.

Keywords Fieldwork • Group work • Leadership development • Solid waste • Thailand • Urban environmental management • Urban water

T. Hoshiko (✉)
Asian Program for Incubation of Environmental Leaders (APIEL),
Department of Urban Engineering, Graduate School of Engineering, The University of Tokyo,
7-3-1 Hongo, Bunkyo-ku, Tokyo 113-8656, Japan
e-mail: hoshiko@env.t.u-tokyo.ac.jp

T. Akiyama
Graduate Program in Sustainability Science, Graduate School of Frontier Sciences,
The University of Tokyo, Environmental Studies Building 334, 5-1-5 Kashiwanoha,
Kashiwa, Chiba 277-8563, Japan
e-mail: akiyama@k.u-tokyo.ac.jp

T. Mino and K. Hanaki (eds.), *Environmental Leadership Capacity Building in Higher Education*, DOI 10.1007/978-4-431-54340-4_4,

4.1 Introduction

Asian developing countries have recently been facing serious urban environmental problems. This is the case especially with large cities undergoing rapid development. Thailand is not an exception and consequently, urban environmental management is a highly important issue. The Thailand Unit is established to foster students' environmental leadership by examining sustainable urban environmental management. The unit dealt with three important issues of urban environmental management: (1) solid waste management (SWM), (2) urban water use and management, and (3) urban flood management in 2009, 2011 and 2012, respectively.[1] Today, disaster-related issues are also becoming increasingly important for sustainability and environmental leadership projects. In fact, the unit held in 2012 focused on the flooding in Bangkok in 2011, which was introduced in the earlier Sect. 2.3.3.

This chapter explores future challenges to improve the design and implementation of the Thailand Unit by examining two cases of the unit conducted in 2009 and 2011 in collaboration with Asian Institute of Technology (AIT) and Kasetsart University (KU). The remainder of this chapter is organized to show how the two cases of the unit were developed, how the programs were implemented year by year, and how the educational effects were examined, as well as review of environmental leadership development in students.

4.2 Development of the Thailand Unit

In this section, background of the unit themes and characteristics of the unit design are described. We compared two different cases of the unit focusing on the core concepts, program content development and educational methods.

4.2.1 Themes of the Unit

The purpose of the unit was to develop in students, as future environmental leaders, a diverse, balanced and integrated understanding of environmental issues. The unit was designed to provide students with holistic and multifaceted information on the unit theme through a comprehensive series of lectures, fieldwork to experience real-world local environmental problems and intensive group discussion which encourages students to share different views, practice consensus building and improve their communication skills. By doing so, the unit tried to enable students to broaden their perspectives and develop on-the-ground competency to identify and resolve environmental problems.

---

[1] In 2010, implementation of the field exercise unit was cancelled and postponed until 2011 due to concern over political unrest in Thailand.

Backgrounds of the two unit themes of urban environmental management are as follows: Management of solid waste involves several direct and indirect issues. A complex problem by nature, SWM takes priority in the agenda of programs on environmental education. While theories on SWM are taught at different levels, a course offering a systemic view is uncommon [1]. The first field exercise "Sustainable Solid Waste Management in Asian Developing Countries (2009)" was structured with the aim of providing hands-on experience in solving real-world waste management problems. A field case in Nonthaburi Province was the highlight of the program. Nonthaburi Province is adjacent to Bangkok and its urban population has been growing accompanying the economic growth and expansion of transportation infrastructure in Bangkok. Although the solid waste has not been collected from Bangkok, the amount of waste has been increasing in the province and there are several environmental problems such as an inefficient system of waste collection, leachate from the landfill, etc.

The second field exercise "Sustainable Urban Water Use and Management in Bangkok (2011)" was structured within the context of a tropical region and the multi-sector dimensions of the issues. Water resource management is a critical issue under both regional and local conditions and especially in tropical regions, it is vulnerable and access to safe water is limited. The focus was centered on Bangkok, where several complex management problems and challenges exist, such as increasing demand of water, an inefficient supply system, and administrative as well as social aspects including economic feasibility of expansion or upgrading of the infrastructure and management system. Both themes were approached by a blend of components of theory, practice, fieldwork and discussion, which is the unique style of the APIEL field exercise.

4.2.2 Concepts and Group Work Task of the Unit

To develop the unit (program), faculty members shared ideas about educational approaches and possible contents and agreed on the following concept in 2009: To develop leadership in students, the most important is for the students to "find" or "identify" issues and problems in the field through discussion among themselves. Namely, we took a "project finding" approach from a real-world experience. Thus, we decided to first bring students to field sites after introductory lectures on basic aspects of the theme, and then lead them in group discussions on the issues actually witnessed in the field. Then there were more lectures and related fieldwork that helped students set and analyze specific tasks of their group work projects, followed by intensive group work to prepare the final presentations.

In 2011, emphasis was made on proposal development for sustainable urban water use and management as well as sustainability indicator analysis to help develop the proposal plan using a scientific approach. Sufficient information on the current important issues of urban water use and management were provided through lectures and field activities. Based on that, students set a future vision and are required to search an effective approach to realize the vision. Namely, we took an "approach

Table 4.1 Group work task

(a) Year 2009	
Approach	Project finding
Instruction	"Based on your own interests, and using the preliminary assignments, lectures and fieldwork during the field program, find a problem to be solved and set concrete objectives for the group work theme. Each group has to frame the work structure to be accomplished by the end of the field exercise, then present conclusions, solutions, and proposals."
(b) Year 2011	
Approach	Approach finding
Instruction	1. Invent sustainability indicators for sustainable urban water use and management
	2. Apply the indicators to measure and evaluate the current situation in Bangkok
	3. Find problems in the current situation using the evaluations
	4. Based on the analysis, come up with solutions for improvement and develop plans for sustainable urban water use and management for the year 2030 in Bangkok

finding" strategy during the proposal on a development plan for 2030. We expected students to have a clearer image to act as environmental leaders in their actual future.

For group work, students were divided into three groups looking for a good balance of the three universities to share their different backgrounds. In line with the project finding approach in 2009, they are instructed to set their own tasks following instructions as shown in Table 4.1a. They are also advised that it is important to frame the group projects with multiple stakeholders and multi-disciplinary solutions in minds.

In 2011, based on the approach finding strategy, students are instructed to develop future plans with scientific and practical approaches that they must find by themselves. Table 4.1b shows the instruction steps. Starting with sustainability indicator analysis to evaluate the current situation of Bangkok, the target year of the proposal plan was set at 2030, a time when we expect the students to be actively contributing to the society as environmental leaders. The results of the group work will be shown in the following results Sect. 4.3.1.

In order to obtain achievements from the unit in terms of academic contents, it was also considered valuable for the students to experience an international conference in the region to present their group work outcomes. In the initial phase of APIEL's educational program development, the Department of Urban Engineering (UE) of the University of Tokyo (UT) had a clear idea that it is important for our field-oriented education to effectively use existing academic networks and resources in Southeast Asia. These networks had been already established when organizing the International Symposium on Southeast Asian Water Environment (SEAWE), which has been successfully run by the department every year since 2003. Therefore, the unit programs were developed to be held in conjunction with the SEAWE.

4.2.3 Content Development of the Unit

Figure 4.1 shows the program contents developed for the unit 2009. The lecture part was arranged to cover components of technology, management, and policy, as well as introduction/fundamentals and looking ahead (at first and fourth stages in Fig. 4.1). We also invited outside experts and policymakers as lecturers. This structure was prepared to cover important issues of solid waste management in a multidimensional and comprehensive way.

Field activities in 2009 were arranged at Sainoi landfill site in Nonthaburi, a waste transfer station, a composting plant, and a medical waste incinerator in Bangkok, an E-waste recycling center at Suan Kaew temple, and a used electric appliance trading market. At the Sainoi landfill site, for example, students can learn about a leachate problem which pollutes surrounding water environments, greenhouse gas (GHG) emission issue and its possible recovery, appropriate application of landfill technology, problems around waste-picker and informal sector involvement, etc. Seeing the real-world local problems, some of which should be solved in a local governance context, and considering that some are influenced by a regional interaction, while others should be considered in a global context, student groups come up with their own projects to undertake.

In 2011, in order to prepare comprehensive information on the program theme, we drew a conceptual framework of the issues of urban water use and management as shown in Fig. 4.2. Important aspects were considered to be water quantity, quality, governance and technology. For the quantity aspect, the introductory lecture covered interactive water demand from three sectors—urban, industry and agriculture, and a field visit to an irrigation project also covered an agricultural water issue. Promotion of 3Rs (reduce, reuse and recycle) for water resources was introduced in a lecture and also a field activity was arranged to observe a wastewater reuse application in a real case. Regarding the quality aspect, field survey on Chao Phraya River water and canal water quality sampling was planned. As for the governance aspect, Thai policies and administration were covered in lectures by local and national government officers, and international governance issues were also covered in another lecture within the tropical region context. Technology aspect was also covered by lectures and various advanced technologies from cases of Tokyo and Singapore were introduced. In addition, climate change impact issues and NGO involvement in water pollution control and economic instruments were incorporated for the analysis of future perspectives.

4.2.4 Educational Methods

Methods of the analysis of educational effects, unit design and leadership development are as follows:

1. In order to examine the educational effects of the field exercise unit in terms of its academic contents, topics of students' group work projects were collected

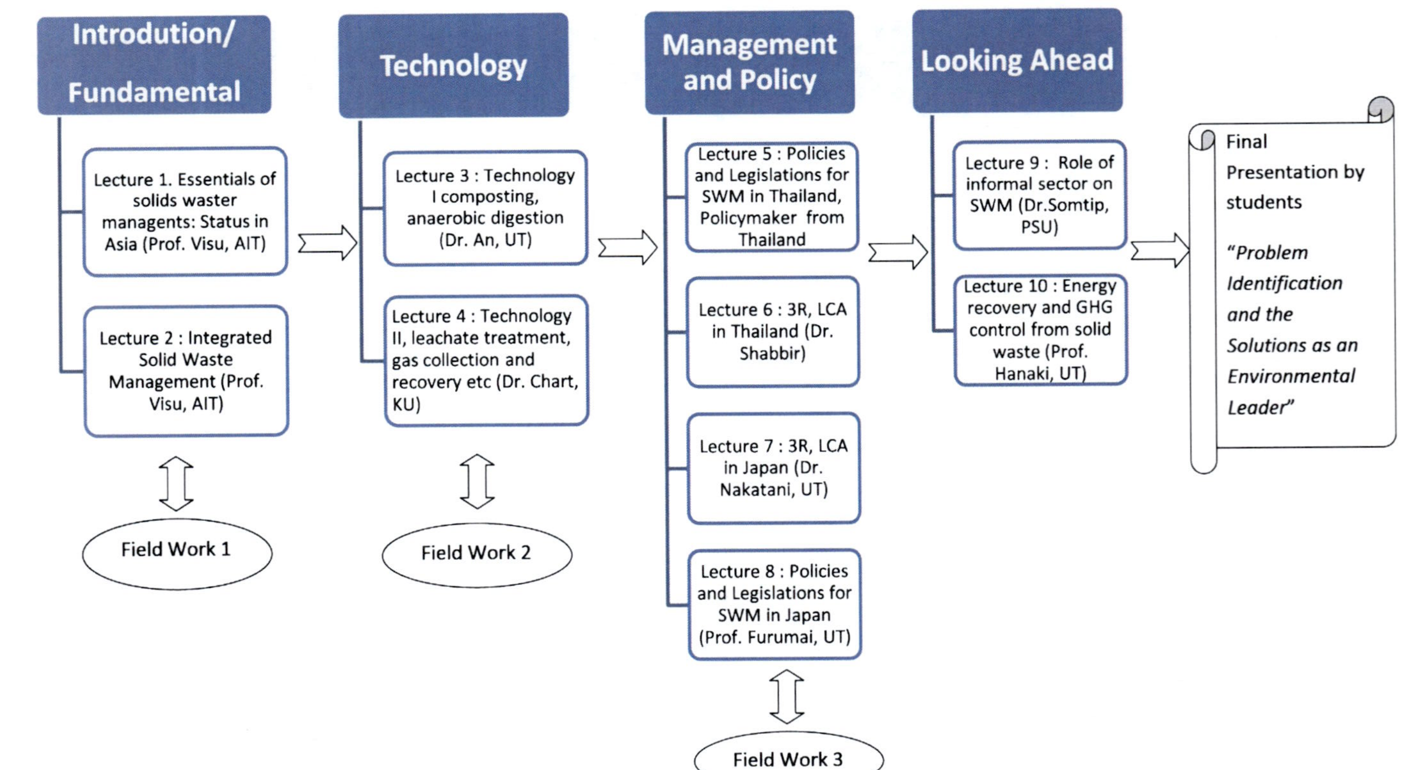

Fig. 4.1 Structure of the lectures of Thailand Unit in 2009

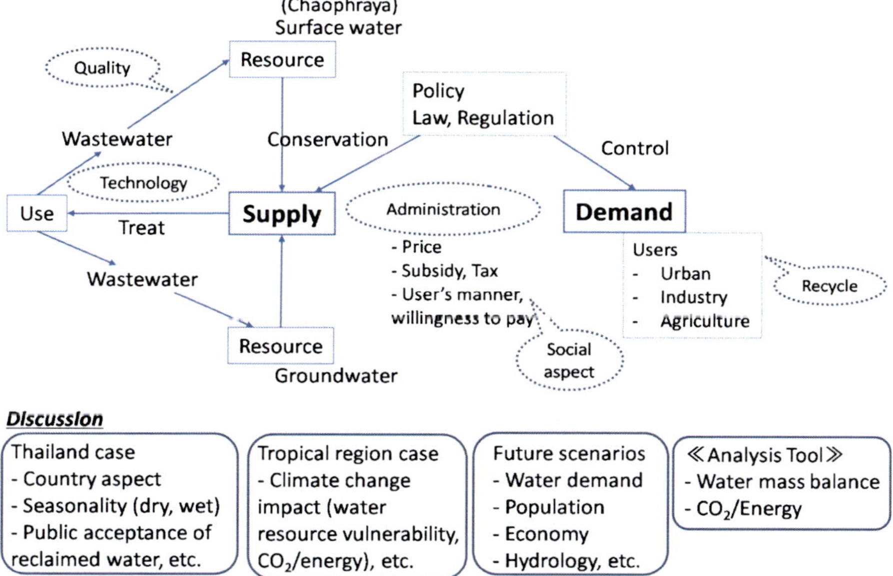

Fig. 4.2 Conceptual framework of the issues of urban water use and management

from the two cases in 2009 and 2011, which employed two different approaches to the group work.

2. In order to improve the unit design and implementation, questionnaire surveys of involved faculty members were conducted to ask strengths and weaknesses.
3. Effects of the unit participation on leadership development in students were examined through feedback comments from the students themselves.

4.3 Implementation of the Thailand Unit

Factual information on the unit implementation is presented in this subsection, including participants' background as well as the field program schedules.

4.3.1 Participants

Participants of the unit in 2009 were six students from UT, six AIT students, and five KU students. In 2011, five UT students, five AIT students, and five KU students participated. Their genders and nationality are shown in Table 4.2.

Academic backgrounds of participants were different in 2009 and 2011. In 2009, the students from Thai counterpart were only from the discipline of environmental engineering. To further diversify the group, in 2011; Thai local counterparts expanded their student backgrounds to come from all the disciplines in AIT and

Table 4.2 Student participants

	UT	AIT	KU
(a) Year 2009			
Number of participants	6	6	5
Gender	4 male, 2 female	3 male, 3 female	1 male, 4 female
Nationality	4 Japanese, 1 Filipino, 1 Bolivian	2 Pakistani, 2 Nepali, 1 German, 1 Thai	5 Thai
(b) Year 2011			
Number of participants	5	5	5
Gender	4 male, 1 female	2 male, 3 female	2 male, 3 female
Nationality	1 Japanese, 2 Chinese, 1 Indian, 1 Nepali	1 Chinese, 1 Nepali, 1 American, 1 Thai, 1 Sri Lankan	5 Thai

from other departments in addition to the Department of Environmental Engineering at KU, which was actually effective in the process of the proposal development in the group work from wider views.

4.3.2 Program Schedule of the Unit

4.3.2.1 Thailand Unit 2009

In 2009, prior to the field program, preliminary assignments were given to the students as follows: a literature review and an exercise on lifecycle assessment (LCA) of solid waste management to learn an analytical tool. One group actually applied this tool for their group project based on this preliminary assignment.

Figure 4.3 shows the program schedule in the field of the Thailand Unit in 2009. The program contained ten lectures, three classroom exercises, four fieldwork trips—Fig. 4.4 shows scenes of the field activities—and group work at the students' initiative almost every day, followed by the symposium and presentations. This unit was held in conjunction with the 7th SEAWE, October 28–30, 2009, at AIT, where student posters were presented on the outcomes of the field exercise. After the field program, summary reports were submitted and wrap-up presentations were given at UT.

4.3.2.2 Thailand Unit 2011

In 2011, prior to the field program, students were given preliminary assignments as follows: a literature review and proposal development exercise for water problems in Asian countries.

The program schedule in the field of the Thailand Unit in 2011 is shown in Fig. 4.5. It included eight lectures, two classroom exercises, and two full days of

	Date	Venue	Morning 9:00-10:30	Morning 10:50-12:20	Afternoon 13:30-15:00	Afternoon 15:20-16:50	Evening
1	21-Oct	KU					Orientation
2	22-Oct Thu	KU	Lecture 1: Fundamentals	Exercise 1: Case work	Fieldwork 1: SWM in Bangkok		Group work
3	23-Oct Fri	AIT	Lecture 2: Fundamentals	Exercise 2: Case work	Fieldwork 2: Sainoi landfill site in Nonthaburi		Group work
4	24-Oct Sat	KU	Lecture 3: Technology	Lecture 4 : Technology	Lecture 5 : Management and Policy - Thailand	Group Work : Setting the theme and framing for the group work	Fieldwork 3: Used electronic appliance trading market
5	25-Oct						
6	26-Oct Mon	KU	Lecture 6: LCA	Lecture 7 : LCA	Exercise 3: LCA	Fieldwork 4 : E-waste recycling activities (Wat Suan Kaew)	Group work
7	27-Oct Tue	KU	Lecture 8: Management and Policy - Japan	Group work: Interim progress report	Lecture 9: Role of informal sector	Lecture 10: Global - Energy recovery and GHG control	Group work
SEAWE Symposium							
8	28-Oct Wed	AIT	Excursion by the Symposium				Group work
9	29-Oct Thu	AIT	Opening ceremony	10:30-12:30 APIEL session	Attendance at the symposium sessions		Group work
10	30-Oct Fri	AIT	Attendance at the symposium sessions		13:00-14:30 Poster presentation	Closing ceremony	16:30-18:00 <APIEL Internal WS> Student final presentation & Closing	UT students go to the airport

Fig. 4.3 Program schedule in the field in 2009. Yellow, blue, green and orange colors indicate time slots for the lecture, classroom exercise, fieldwork and group work, respectively

Fig. 4.4 Scenes of field activities in 2009. (**a**) Interview survey at the recycling market. (**b**) Waste transfer station. (**c**) Leachate test at the Sainoi landfill site. (**d**) Sainoi landfill site

fieldwork and group work at the students' initiative almost every day. Figure 4.6 shows scenes of group work and on-the-field activities. Since the symposium presentation on the group work outcomes was separate from the field program schedule this time, the schedule was less packed than that in 2009, and we were able to give the students more time for discussion. After the program in the field, summary reports were submitted and wrap-up presentations were conducted at UT. Afterward, the outcomes of the group work were presented as posters in the 9th SEAWE held in Bangkok on December 1–3, 2011.

4.4 Results and Reviews of the Thailand Unit

4.4.1 Group Work Results

To examine the educational effects of the unit programs in terms of their academic contents, the group work results are presented in this subsection. Details are summarized in Table 4.3 for the case in 2009 and in Table 4.4 for the case in 2011.

	Date	Venue	Morning 9:00–10:30	Morning 10:50–12:20	Afternoon 13:30–15:00	Afternoon 15:20–16:50	Evening
0	18-Aug	AIT	UT students arrive at the BKK airport and move to AITCC				
1	19-Aug Fri	AIT	Opening & Introduction	Lecture 1: Wastewater reuse: Urban, agriculture and industry sector interactions	Lecture 2: Challenge to sustainable urban water use in Tokyo	Exercse 1: Case study	Gathering event at AITCC
2	20-Aug Sat	AIT	Exercse 2: HW presentation, Brainsorming session	Lecture 3: Environmental leadership			
3	21-Aug	AIT					
4	22-Aug Mon	AIT	Group Work	Lecture 4: Social study on water governance in Thailand and international relations	Lecture 5: Perspectives on NGO involvement in water pollution control and economic instruments	Lecture 6: Singapore case cn urban and industrial wastewater reuse	Group work
5	23-Aug Tue	Field	Fieldwork 1: Lecture at BMA, Wastewater reuse application, Rattanakosin WWTP, Water quality sampling at Chao Phraya river and canals				Group work
6	24-Aug Wed	AIT	Lecture 7: Integrated Water Environment Management in Thailand		Lecture 8: Water reuse technology	Group work : Interim progress report	Group work
7	25-Aug Thu	Field	Fieldwork 2: Lecture at JICA, Irrigation projects, Wastewater reuse in agriculture				Group work
8	26-Aug Fri	AIT	Group work				Group work
9	27-Aug Sat	AIT	Final presentation	Closing Ceremony			
	30-Aug		UT students return to Tokyo				

Fig. 4.5 Program schedule in the field in 2011. Yellow, blue, green and orange colors indicate time slots for the lecture, classroom exercise, fieldwork and group work, respectively

Fig. 4.6 Scenes of field activities and group work in 2011. (**a**) Water sampling of Chao Phraya River. (**b**) Brainstorming session. (**c**) Agricultural field visit. (**d**) Water quality measurement of Chao Phraya River

4.4.1.1 Thailand Unit 2009

In 2009, group work was conducted based on the project finding approach, which involved intensive discussions on topic finding and consensus building under strict time limits. Because of this approach, more than one student commented afterward that consensus building at the initial phase was one of the most difficult and time-consuming parts, since each person had different ideas and would not easily be persuaded to change them. There were also language barriers, and students encountered different attitudes, dependent on culture and communication style. One student also said that she understood that with limited resources the real situation and problems so complex that it becomes difficult to prioritize the problems. From the presented results, however, it was pointed out that the given group work tasks were sufficiently met and solutions for the addressed problems were clearly proposed. Thus, consensus was made, language barriers were not fatal, and prioritization of the problems was achieved. Furthermore, it was revealed that field visits with support of comprehensive lectures were very good materials, in combination with the project finding approach, to learn practical issues and to do training on problem solving in the context of a team.

These results were presented in the 7th SEAWE. It was incorporated at the end of the field program schedule, which was the cause of strict time limits for the students to complete their projects. The presentations were successfully done through

Table 4.3 Group work results (2009)

Group 1	**Title: Sustainable E-waste management in developing countries—economic benefits and health risks [2]**

Abstract: Focusing on the health issue of solid waste, proper E-waste management systems in developing countries were discussed. The objectives were to propose a safe E-waste recycling system at the local level with lower risk for workers and the environment, especially using case studies in China and Thailand, and to show their advantages and disadvantages. Health risks and environmental impact in China and successful E-waste management at the Suan Kaew temple in Nonthaburi were reviewed. E-waste recycling creates job opportunities in developing countries, saves resources, and at the same time gives the underprivileged access to electric and electronic equipment. However, most people there work under poor conditions due to a lack of awareness and understanding of occupational health regulations. The institutionalization of E-waste management, including training on health risks and guidelines for adequate working conditions, can help overcome the negative impact.

Group 2	**Title: Sustainable vision for SWM in Bangkok [3]**

Abstract: Through the field visits, problems were identified in improper waste management in each unit operation resulting unsanitary conditions and low quality of life (QOL). The objective was to develop a vision for sustainable SWM aiming at overall improvement of the whole management system in Bangkok through intensive discussion on how to achieve the vision at this transition stage—from the current situation towards the goal. According to the analysis of waste flow in Bangkok, it turned out that there is strong potential to improve all the steps of SWM. Big problems include an ineffective infrastructure and lack of awareness and knowledge by workers and householders. Cooperation with the public and private sectors is also needed. Furthermore, the "informal sector" has great potential to do recycling in more effective ways as a business. We know that a developing country can't change quickly and that this requires step-by-step improvements. So, we will rethink the current system, improve existing facilities, try to reorganize the whole system and arrive at a future vision. Involving local people makes the system more practical in real society. We have to consider not only the management and technical problems but also the social systems. This should be the most effective way for proper, integrated SWM.

Group 3	**Title: GHG emission reduction potential in a solid waste disposal site—a case study of the Sainoi landfill [4]**

Abstract: Reducing GHG (greenhouse gas) emissions from the landfill was the main challenge. The objectives were to estimate baseline GHG emissions from the landfill site and to estimate emission reduction by considering three scenarios: flaring, generating electricity from captured biogas, and incineration using a LCA approach. The findings were that CH_4 accounts for 84% of all of the GHG emissions. Recovery of CH_4 would have a large effect on GHG reduction (75%). Therefore, methane gas collection is an important factor [technique]. CH_4 flaring would reduce GHG emissions by 22%. The GHG emission potential for electricity generation in Thailand is larger than Japan. Therefore Thailand has more incentive to use landfill gas to generate electricity compared with Japan. Incineration can reduce up to 75% of GHG emissions. If incineration is introduced, it emits a greater amount of N_2O, so technology for reducing N_2O emissions is also important. If we used a factor in Thailand especially for CH_4 emissions from landfills, the result would be altered. In such a case, incentives for introducing CH_4 collection or incineration systems would be large. Based on the LCA analysis results, a scenario for the recovery and use of landfill gas for generating electricity had the largest emission reduction potential.

Table 4.4 Group work results (2011)

Group 1	**Title: A framework for analysis of wastewater management system in Bangkok metropolitan area using sustainability indicators [5]**

Abstract: In order to deal with wastewater, safeguard public health, and protect the natural environment in a sustainable way, a framework for analysis of the management system using sustainability indicators is needed to evaluate and improve the current system. The wastewater management system mainly includes wastewater collection and treatment. Sustainability indicators were used to consider the environmental, economic and social factors—employing the Drivers–Pressure–State–Impact–Response (DPSIR) methodology—for the seven existing wastewater treatment plants and the entire wastewater management system in Bangkok. Among the three factors (above), only the investment item in the economic factors was positive. The other items were evaluated as negative and far from satisfactory for sustainable operation. The current bad water quality is also threatening public health and natural sustainability. Finding problems is easier than solving them. Both Bangkok's government and its citizens face big challenges: How to improve people's awareness to change their own lifestyle; how to sustain and manage the wastewater treatment plants; how to design treated water reuse plans; how to improve treatment technology with less energy (less money); how to popularize and deepen sustainable education; and how to create more job opportunities and improve personal incomes together with the increasing need for higher living standards and the needs of the surrounding environment.

Group 2	**Title: Challenges and opportunities for achieving sustainable urban water use and management in Bangkok 2030: proposal for sustainability indicators [6]**

Abstract: The goal is to define sustainability indicators for urban water use. Managing the quantification of that sustainability using multidimensional indicators is a complicated issue. In Bangkok, annual water demand is growing at 8 %. At the same time, the quantity of wastewater has been increasing at a much faster rate. Currently only about 50 % (at a maximum) of the total wastewater is treated in Bangkok. As well, there are other problems that urgently require improvement, such as a fresh water shortage, and flooding and deterioration of water quality in the dry season. Not only technical issues but also socio-cultural issues must be considered in greater detail. Public acceptance of reusing reclaimed water is a major concern, no matter how good the quality. Effective communication among all the stakeholders—mainly from the government, public, industrial, and economic sectors—is crucial in setting up sustainable urban water use and management. Based on a review of these current problems, the following suggestions were made: a commitment at the policy level for sustainable urban water use and management, improvements in wastewater collection, promotion of energy and water-saving technologies, and the promotion of using reclaimed water.

Group 3	**Title: Improving Quality of Life (QOL) for Bangkok's citizen through sustainable water use and management [7]**

Abstract: Bangkok is one of the urban and economic mega-centers in Asia. But the prosperity of Bangkok comes with urban hurdles, especially infrastructure and environmental weaknesses. The increasing gap between supply and demand is posing a greater threat and the city is presently facing multiple threats to its water environment, including deteriorating water quality (ground and surface water), water accessibility, wastewater management, water governance, saltwater intrusion, etc. Discussions were held that identified indicators, grouped into three categories; water and wastewater management, integrated water management, and governance. Discussions were also held to find the correlation between urban poverty and water environment, and to find out the influence of the degrading water environment on the quality of life (QOL) of Bangkok's citizens. Taking QOL as the representative indicator, the objective of the proposal development was to improve QOL for Bangkok's citizens by creating a mechanism for sustainable water use and management. The physical, social, and environmental dimensions of QOL were investigated, and the final outcome was formulated with both short-term and long-term holistic recommendations towards 2030 for the categories of water and wastewater management, as well as integrated water management and governance.

their efforts. The connection between the contents of the unit program and the group work topics are as follows: Group 1 picked a topic from the field visit to the E-waste recycling activity at Suan Kaew temple, and addressed the related health issues in their project. Group 2 took a hint from the field visit to the waste transfer station in Bangkok, where they observed several improper waste management practices and addressed the need for a vision for sustainable SWM. Group 3 took the example of Sainoi Landfill site in Nonthaburi and addressed GHG emission reduction issues by applying technological scenario analysis using a LCA approach.

4.4.1.2 Thailand Unit 2011

In 2011, group work was conducted based on the "approach finding" strategy for the proposal development. Given sufficient information on the unit theme and relatively less pressure of time restrictions, students were able to spend time on discussions about the sustainability indicator analysis and on the proposal development. Because the importance of the scientific approach for the proposal development was emphasized, students worked hard to collect comprehensive information by themselves and to find a logical approach to develop a proposal plan for their vision for 2030. In other words, this approach helped students develop practical leadership based on wider knowledge and expertise about the concerning issues, instead of having only idealistic but vague images for the future.

Table 4.4 shows details of the group work outcomes, which were presented in the 9th SEAWE, more than 3 months after the unit implementation. The logical approaches taken by the student groups are explained as follows: Group 1 employed an approach of system thinking and framework analysis to systematically explain their plans for the sustainable urban water use and management towards 2030. Group 2 focused on sustainability indicator development to show their visions for the 2030 qualitatively and quantitatively. Group 3 highlighted QOL as one of the most important indicators and proposal plans were presented to improve the QOL of Bangkok citizens.

4.4.2 Reviews of the Unit

To evaluate the program design and unit performance, feedbacks from the involved faculty members are shown, clearly indicating strengths and weaknesses or points for improvement in Table 4.5. There was a common strength of the unit design for both cases in that the unit was conducted by a nontraditional style of teaching with a blend of theory, practice, fieldwork and discussion. Two different approaches of the group work were also evaluated as strengths, as they work effectively to achieve the group work task. Overall, the first case in 2009 had more weaknesses and the second case in 2011 had more strengths, which indicates significant improvements in the unit design and implementation as the faculty's experience and teaching capacity increased.

Table 4.5 Summary of the review on the two cases

	Year 2009	Year 2011
Strengths	– Nontraditional style of teaching with a blend of the theory, practice, fieldwork and discussion. – "Project finding" approach was effective in seeing a real-world problem with a critical eye and in practicing consensus building. – Group work outcomes were presented at the SEAWE and all the participants could attend the SEAWE, because it was incorporated in the field program.	– Nontraditional style of teaching with a blend of the theory, practice, fieldwork and discussion. – "Approach finding" strategy was effective in developing realistic leadership based on wider knowledge and expertise about a concerning issue. – Improvement of the several weaknesses pointed out in 2009, including better information distribution, longer Q&A time in the lectures, which enhanced interaction between the lectures and students, less pressure of time limitation on students for group project completion, sufficient time for discussion and for the preparation for SEAWE presentation. – Group work outcomes were presented at the SEAWE.
Weaknesses	– Need for better information/course materials distribution for students – Need for better course management, e.g., more interaction between lecturers and students and among students – Work load was a little too heavy for the students – Despite the good opportunity to coordinate with the symposium, there was a need for enough time between the field program and the symposium presentation for better development of academic results	– More time should be allocated for field activities – More demands on Japanese students from the Thai counterparts

4.4.3 Environmental Leadership Development in Students

To examine the effects of the unit on the leadership development in individual students, their feedback comments are summarized in Table 4.6. From the comments, it is observed that the unit had a large impact on their attitude toward leadership development. The impact varied from one student to the other. Setting the clear short-term goal to encourage SEAWE conference presentations on the results of the field exercise was also effective to boost their motivation and ability to produce concrete outcomes as well as for high-level academic communication.

During the group work projects, many students showed frustration, becoming aware of their current limitations to solve real-world problems and a need for

Table 4.6 Summary of feedback comments on the leadership development

---

Year 2009

"Specific knowledge by itself doesn't change the situation. Knowledge, vision, and communication skills are all needed to solve real-world problems."

"The group dilemma is a microcosm of what is happening in the real world. Policymakers spend too much time debating what should be done, leaving little time for how it will be done. Building consensus, therefore, is a skill that an environmental leader should possess in order to get things done."

"There is an inconsistency in waste segregation in Thailand, which is not unique to the country. The Thai government promotes segregation at the source. However, waste collectors just mixed the waste. It is therefore necessary for environmental leaders to conduct training and education for workers and the public so that they will better able to understand and share the same goal as management."

"For environmental leadership development, it is necessary to have a healthy attitude and enthusiasm to try to understand other technologies, cultures, and points of view."

Year 2011

"Things that worked effectively for leadership skill improvement were visualization of your concept, your attitude to initiate for the progress, such as action plan formulation and promotion of mutual understandings."

"In each group, from time to time, some naturally leading people were observed, who were equipped with the power to convince other group members and set clear steps first and head for a consensus on the framework building of the topic. In such an environment, humanity was important for inclusion of all the members."

"Through the experience of the field exercise, the leadership and solution process was thought out and organized to be like the following:

Learn the problem → Visualize the concept and theme → Develop a clear vision → Discuss with team of experts and various stakeholders → Develop a consensus → Reach out to the masses for their feedback → Modify the outcome using the feedback from the masses → Implementation."

"Even when consensus building was difficult among group members, patience, the ability to listen to others, an analytical mind and tenacity were there with everyone to convey their thoughts and listen to others for the best possible outcome. Many of the essential leadership skills seemed to be already in place with the participants but in a sporadic fashion. The success of the program was that it was able to gather the sporadic skills in a constructive way and helped everyone to augment their skills and overcome their deficiencies."

---

improving their knowledge and communication skills, such as better consensus building, facilitation and English skills, etc. Nonetheless, the frustration itself should be considered one of achievements. They experienced the complexity of real-world environmental issues and tried to approach them in new ways, which would have improved diverse, balanced and integrated understanding. This is what we had aimed as the most important purpose of the leadership program, and in that sense, the Thailand Unit's educational challenge bore fruit.

4.5 Concluding Remarks

This chapter showed how the Thailand Unit was developed and implemented year by year, by comparing the two cases on sustainable solid waste management in 2009 and sustainable urban water use and management in 2011. Successful group work results were shown, which verified educational effects of the unit in terms of academic contents. In 2009, project finding approach was taken and it contributed to improvement of consensus building skills, while in 2011, approach finding strategy was taken and it contributed to improvement of logical approaches for the proposal development. Unit design and implementation performance were also evaluated using feedback from the faculty members. According to this evaluation, significant improvement was observed in the second year including sufficient time for group discussion at the students' initiative. Students' feedback comments also showed that the unit had a large impact in environmental leadership development. The purpose of the unit was certainly met, which aimed at achieving diverse, balanced and integrated understanding on environmental issues in students as future leaders, by the educational challenge of the unit and through the students' own great efforts.

Acknowledgments The author would like to deeply thank Prof. Chettiyappan Visvanathan from the Environmental Engineering and Management, School of Environment, Resources and Development, AIT and Dr. Chart Chiemchaisri from the Department of Environmental Engineering, Faculty of Engineering, KU for their cooperation and strong support for the Thailand Unit. Last but not least, the unit was successfully conducted by the initiative and representation by Prof. Hiroaki Furumai from the Graduate School of Engineering, The University of Tokyo.

References

1. Visvanathan C (2009) Course material of APIEL Nonthaburi unit 2009
2. Sano S, Takahashi Y, Chitapornpan S, Theepharaksapan S, Nawaz R, Clausen A (2009) Sustainable E-waste management in developing countries: economic benefits and health risks. Poster presentation at the 7th international symposium on southeast Asian water environment, AIT, Thailand, 28–30 October 2009
3. Sensai P, Muenmee S, Rukapan W, Aoki E, Bellido JCV, Zeshan S (2009) Sustainable vision for solid waste management in Bangkok. Poster presentation at the 7th international symposium on southeast Asian water environment, AIT, Thailand, 28–30 October 2009
4. Baloch AA, Ferrer J, Takahashi N, Naroiet P, Shakya S (2009) GHG emission reduction potential in a solid waste disposal site: a case study of Sainoi landfill. Poster presentation at the 7th international symposium on southeast Asian water environment, AIT, Thailand, 28–30 October 2009
5. Ghimire A, Franklin K, Chen S, Boonnorat J, Prommetjit P, An K, Hoshiko T, Honda R (2011) A framework for analysis of wastewater management system in Bangkok metropolitan area using sustainability indicators. In: Proceedings of the 9th international symposium on southeast Asian water environment, pp 125–129
6. Khanal R, Chu C, Kalaimathy SN, Thiamngoren P, Patchanee N, An K, Hoshiko T Honda R (2011) Sustainable urban water use and management: scenario of Bangkok 2030. In: Proceedings of the 9th international symposium on southeast Asian water environment, pp 120–124
7. Biswas A, Boonyaroj V, Denpetkul T, Zhirong H, Kobayashi H, An K, Hoshiko T, Honda R (2011) Improving quality of life of Bangkok's citizen through sustainable water use and management: vision 2030. In: Proceedings of the 9th international symposium on southeast Asian water environment, pp 103–108

Chapter 5
Environmental Leadership Education for Tackling Water Environmental Issues in Arid Regions

Tomohiro Akiyama and Jia Li

Abstract This chapter introduces one of the APIEL field exercises, the Oasis Unit, which is conducted in northwestern China. To equip the students with a wide knowledge base and practical skills, this unit is strongly field-oriented and applies in its course design the Integral Approach proposed by Ken Wilber. The approach provides a trans-/cross-disciplinary framework for identifying environmental problems of complexity, as well as bringing together methodologies from different fields and leadership qualities. After four years of implementation, the approach is considered successful in educational program design for environmental leadership and for promoting the leadership development of participants.

Keywords Environmental leadership education • Field exercise • Integral Approach • The Heihe River basin

5.1 Introduction

The United Nations Decade of Education for Sustainable Development, initiated in 2005, aims to develop and implement educational programs that focus on the three pillars of sustainability, i.e., environment, economy, and society. APIEL

This chapter is an updated version of Akiyama et al. [25].

T. Akiyama (✉)
Graduate Program in Sustainability Science, Graduate School of Frontier Sciences,
The University of Tokyo, Environmental Studies Building 334,
5-1-5 Kashiwanoha, Kashiwa, Chiba 277-8563, Japan
e-mail: akiyama@k.u-tokyo.ac.jp

J. Li
Faculty of International Studies and Regional Development, University of Niigata Prefecture,
Ebigase 471, Higashi-ku, Niigata, Niigata 950-8680, Japan
e-mail: lijia@unii.ac.jp

T. Mino and K. Hanaki (eds.), *Environmental Leadership Capacity Building in Higher Education*, DOI 10.1007/978-4-431-54340-4_5,
© The Author(s) 2013

was established in line with the Japanese government's initiative to promote sustainability education in higher education institutions and to nurture environmental leaders. Since the concept of sustainability and present-day environmental problems are featured by the complexity of issues, APIEL has paid considerable attention to trans-disciplinary and/or cross-disciplinary education. In particular, APIEL focuses on fostering student leadership through on-site curriculum development.

At the core of the on-site curriculum, there is a course entitled as "Field Exercise" built for the purpose of practical learning. To guarantee the diversity of education methodologies and educational effects, APIEL's field exercise units can be roughly classified as either field-oriented or structure-oriented. In either case, the course consists roughly of four parts: preliminary learning; on-site learning; after-the-fact learning; and joint tasks (report preparation and presentations). Field-oriented exercises develop environmental leadership through group-based work led by students that relates to preliminary studies, planning of research activities in the field, implementation of field surveys, and the completion of solution proposals. Akiyama et al. [1] mentioned that, field-oriented exercise fosters environmental leadership mainly through the execution of field surveys, while the structure-oriented exercise fosters environmental leadership mainly through a series of educator-structured programs (i.e., in-class lectures, discussions, and short field trips).

This chapter will provide a detailed description of one of the field-oriented exercises, the Oasis Unit, which has been conducted in the Heihe River basin in arid northwestern China. The Oasis Unit especially emphasizes on the multiplicity of sustainability education and necessity for a holistic view to understand the various dimensions of environmental issues. The unit, therefore, makes the effort to apply the Integral Approach proposed by the American philosopher Ken Wilber into the practice of environmental leadership education. In this chapter, we provide an introduction to this approach, followed by its benefits for integrating academic disciplines, as well as the organizational resources and the perspectives of stakeholders used for finding solutions to environmental issues. This chapter concludes by identifying future challenges of field-oriented environmental leadership education.

5.2 Integral Approach: A Simplified Introduction

The Integral Approach aims to incorporate multiple perspectives from around the world instead of focusing on specific objects and/or specific systems of objects. Given the complexity of reality, the Integral Approach cuts across fields and brings together existing methodologies into a trans-/cross-disciplinary framework. According to Wilber [2, 3] all phenomena in the world can be categorized into four groups using a four quadrant framework. These quadrants are four ways of viewing the same occurrence in the reality from four different perspectives.

They are located in the interior and exterior of both individuals and collectives. The exterior aspects are found on the right-hand side, with physical and behavioral aspects in the upper right quadrant and social systemic aspects in the lower right quadrant. The interior aspects are found on the left-hand side, with intentional, personal, and psychological aspects in the upper left quadrant and cultural aspects (collective values) in the lower left quadrant. Although the four quadrants are ontologically distinct, there is nevertheless an interwoven, intimate correspondence among them.

Wilber's Integral Approach has received substantial attention and has been applied to a variety of fields, in both academia and practice. In particular, it is gaining attention around the world from researchers and practitioners in the field of sustainability/environment. We apply this approach because, first, it provides a holistic framework encompassing a wide knowledge base in the social, economic, cultural, and natural sciences, and second, it leads to the successful implementation of an environmental leadership education program by providing a comprehensible structure for educational curriculum design.

In the academic world of sustainability/environment, we especially draw on the following studies: Esbjorn-Hargens and Zimmerman [4], Eddy [5], Kayane et al. [6], Kayane [7, 8], Esbjorn-Hargens [9], Voros [10], and Floyd and Zubevich [11]. Among these, Kayane et al. [6] and Kayane [7, 8] are the pioneering studies that applied the approach to water environment issues. They analyzed the water environment and related changes in Lijiang City, China and Tsuwano Town, Shimane Prefecture, Japan. They argued that, first, the natural environment, especially water, is a common element related to all quadrants, and second, that the current environmental problems are often consequences induced by the abnormal development (evolution) of the lower right quadrant, i.e., rapid technological innovation in the twentieth century. We mention their works because the main topic of the Oasis Unit is water scarcity and water resources management in arid regions. The framework outlined in Kayane et al. [6] and Kayane [7, 8] was further developed in Akiyama et al. [12]. Figure 5.1 is a simplified version of Akiyama et al. [12]'s four quadrant framework related to water environmental issues.

Our framework allocates perspectives on water environmental issues into four associated quadrants. It draws on the conventional concept of sustainability/environment studies, with its emphasis on empirical research methods (quantitative and scientific), as well as alternative concepts, to encompass inter-subjective and subjective modes of inquiry (qualitative, hermeneutic, and introspective). The benefit of this framework, although requiring further research, is profound. On the one hand, it offers a common foundation for people to view various perspectives on the complexity of water environmental issues. In other words, it incorporates knowledge and methodologies from multiple disciplines. At the practical level, as far as we know, Wilber's Integral Approach has previously been applied to the fields of international development and education. In the field of international development, several international development organizations and non-governmental organizations, including UNDP (a global leadership development program around HIV/AIDS), are increasingly seeing the

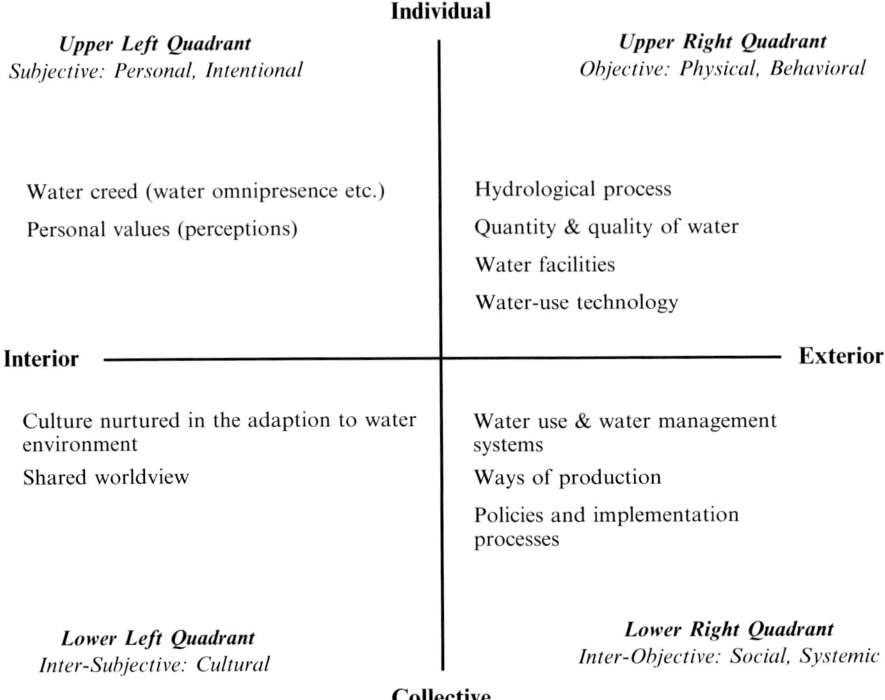

Fig. 5.1 An integral framework for water environmental issues

advantages of adopting the approach to their projects [13]. At the local level, this approach has also been used in community development projects such as the one in the San Juan del Gozo community in El Salvador [14]. In the field of education, the approach has also proved useful for curriculum development: see, for example, Gidleya and Hampson [15], Lloyd [16] and Akiyama et al. [1]. Drawing upon these practices, we designed the field exercise following the four quadrant framework presented in Fig. 5.2.

In Fig. 5.2, "I" (or "we") refer to the participant(s) in the field exercise. This framework helps us to design a field exercise. First, it requires the field exercise design to foster self-development through personal learning as well as group work, collaboration and communication with the different stakeholders. Second, it requires the field exercise design to lead to the common conclusions of all participants, as well as to accommodate their individual views. Therefore, the field exercise should provide enough time and resources for the participants to reach a consensus and to set clear shared goals from the beginning, while also allowing for individual points of view.

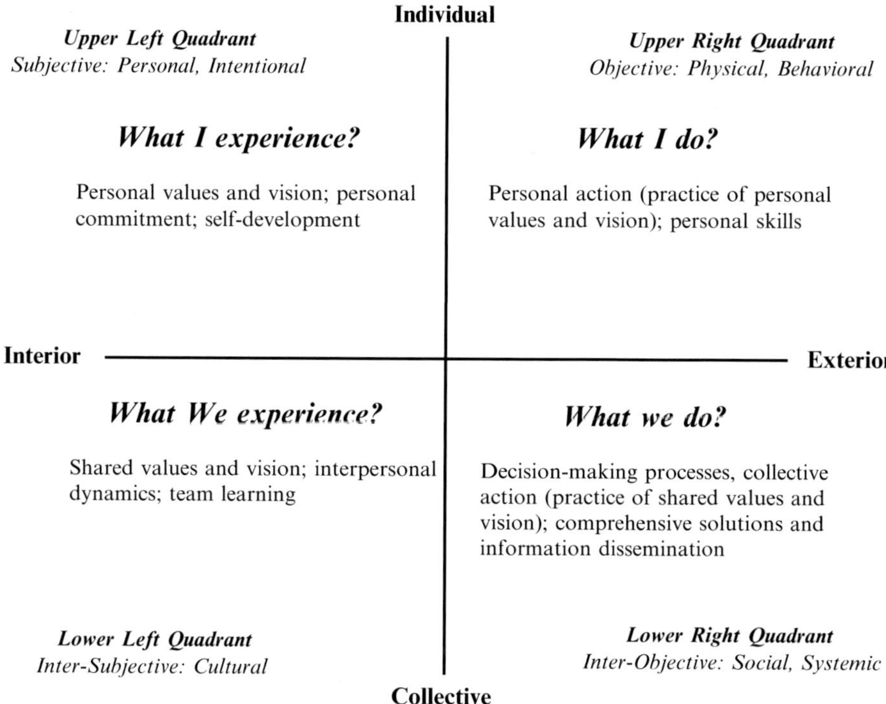

Fig. 5.2 An integral framework for environmental leadership education

5.3   Applying Integral Approach to Environmental Leadership Education

5.3.1   Oasis Unit in Northwestern Arid China

The Heihe River basin in arid northwestern China is an excellent area for fieldwork participants to consider how sustainable development could be achieved in dryland regions under severe water resources constraints. APIEL chose this river basin specifically as a target area for field exercise for the following two reasons.

First, sustainable development in dryland regions is an inevitable, current world challenge. It is associated with water security as well as food security around the world. Today, irrigated agricultural land makes up less than one-fifth of the total cultivated area in the world but produces about two-fifths of the world's food [17]. Irrigation farming, to a great extent, contributed to the increase in food production in the twentieth century and continues to support large numbers in an increasing population. However, food production relying on the "irrigation miracle" gives significant impacts on water resources. Agricultural water use, including irrigation, accounts for about 70% of global water withdrawals [18]. In dryland regions, large-scale development of irrigation farming induces dramatic increase of water demand. Consequently, it often results in stoppage of river flows, dry-up of lakes, decline of groundwater table and other related ecosystem degradation.

Second, the Heihe River basin, the second largest inland river in China, provides many topics for the study of sustainable development in dryland regions. In the Oasis Unit, we highlight several of these: watershed management, water-saving policies (decision-making processes, implementation and assessment), as well as environmental degradation and recovery.

In the Heihe River basin, historically, people living in the middle reaches and the ones living in the lower reaches had different ways of production. People living in the middle reaches adopted irrigation farming (settled culture); while the people living in the lower reaches adopted nomadic husbandry. Since the 1950s, intensive agricultural practices in the middle reaches have resulted in a dramatic degradation of the environment in the lower reaches. Conflicts over water use between the people living in the middle reaches and those living in the lower reaches date back at least 200 years ago [19–21]. However, these conflicts have never been as fierce as today. The intensive exploitation of water resources in the middle reaches has largely declined the amount of water flows to the lower reaches. By 2002, more than 30 tributaries of the Heihe River basin had dried up. In the lower reaches, two terminal lakes dried up in 1961 and 1992, respectively. Riparian vegetation degraded. Salinization and desertification intensified. The desertification in the lower reaches has attracted substantial attention nationwide and is thought to be the origin of dust storms in the spring.

In recent years, a range of environmental conservation activities has been carried out in the river basin, particularly Zhangye, a city in the middle reaches. The main purpose of environmental conservation activities is to preserve the environment in the lower reaches. At the core of those activities is the Integrated Water Resources Management Plan of the Heihe River Basin promulgated by the Chinese State Council in 2001. This plan states that "when the river discharge from the upper reaches amounts to 1.58 billion m3/a, Zhangye City, located in the middle reaches of the Heihe River basin, has to increase discharge of 0.225 billion m3/a to the lower reaches, which means 0.95 billion m3/a should be released to the lower reaches" [22]. In other words, the central government requires the city of Zhangye to reduce water consumption by administrative order. Since 2001, Zhangye has been repeatedly selected as an experimental site for pilot programs of water resource management. In particular, in early 2002, the Ministry of Water Resources of China initiated an experimental project for establishing a water-saving society in the middle reaches at Zhangye. The project was set to save water and increase water use efficiency mainly in two ways: (1) by building concrete irrigation channels using government funds; (2) by introducing market mechanism. The policies include introduction of meters to charge for irrigation water based on the amount used, and the introduction of water use rights system with tradable water quotas. At the same time, in Ejina in the lower reaches, a relocation policy has been implemented because overgrazing was considered one reason for environmental degradation.

5.3.2 Making the Field Exercise Unit Integral

Applying Integral Approach to environmental leadership education is an evolving process that is far from completion. There are external constraints, such as those on human resources, finance and time that prevent the ideal development of a program. In addition, the students, who have come through a relatively narrow education system, do not always know how to respond to a new, holistic way of learning. Therefore, when we design the Oasis Unit, we focus on an integral knowledge base as well as integral practices.

Table 5.1 presents a brief description of the Oasis Unit. Started in 2009, it takes place once a year. With integral thinking as the general framework for program design, we extended the content of the field exercise (from 2009 to 2012) to incorporate more perspectives related to environmental issues, and provided more experiences for students to develop practical skills. The field exercise is jointly organized by APIEL, The University of Tokyo, and Cold and Arid Regions Environmental and Engineering Research Institute (CAREERI), Chinese Academy of Sciences. Students who join the field exercise are from both institutes. They come from different countries and major in several academic fields. To provide the students multi-disciplinary knowledge and multiple views about local environmental problems, faculty members from different academic fields as well as local stakeholders were involved in the different stages of the fieldwork. We have established close relationships with CAREERI and the local water authority to move beyond the limits of universities as well as to let students know that they are tackling real-world problems. The students are required to make policy recommendations and deliver this information to the local water authority. In addition, in 2011 and 2012, the collaboration was strengthened by working with several other institutions from both Japan and China. We hold international symposia in Japan and in China to build a platform for students to hear fresh voices from academia beyond faculty members, government officials, and businesspeople.

Figure 5.3 is an overview of the organizational framework used in our field exercise. Note that students are the leading players. We simply created the space for students to see real-world environmental problems and to realize their own development. In Fig. 5.3, environmental issues (*Issues addressed*) are the research topics covered by the students; *methodologies* are those adopted by the students; *competencies* are the capabilities and/or skills that students are expected to develop through participating in the field exercise.

Problem-solving based learning is the core concept of the course design. It reveals related issues, brings together the necessary research methodologies, and consequently improves participants' competence to become environmental leaders in the future. The main objective of the field exercise is to enhance the students' practical skills through solving specific environmental problems in the real world. Issues in each quadrant have different perspectives for the same environmental problem: sustainable development of the Heihe River basin, which is facing severe water shortages. The issues are interwoven. To provide comprehensive solutions

Table 5.1 Description of the field exercise unit in the Heihe River Basin

	2009	2010	2011	2012
Place(s)	Zhangye, Gansu Province	Middle and lower reaches of the Heihe River basin (Zhangye, Gansu Province, and Ejina, Inner Mongolia)	Zhangye, Gansu Province	Zhangye, Gansu Province
Duration	9 days (August 7–15)	14 days (August 10–23)	13 days (August 27 –September 7)	13 days (August 4–16)
Collaborating institution(s)	CAREERI	CAREERI	CAREERI	CAREERI; Sophia University
Students	9 students from 4 countries	16 students from 7 countries	10 students from 6 countries	12 students from 6 countries
Major subjects of students	Sustainability science, urban engineering, geography	Sustainability science, urban engineering, geography	Sustainability science, urban engineering, geography	Sustainability science, urban engineering, geography, global environmental studies
Academic specialties of faculty members	Seven faculty members: water environmental engineering, hydrology, geology, limnology, geography, economics, and sustainability science			
Stakeholders	Researchers (local and foreign); local government officials (water management authority); local farmers			
Activities	(1) Lectures; (2) site visits; (3) discussions and communications with stakeholders; (4) quantitative and qualitative analyses; (5) group work; (6) results reporting (group-based)			
Follow-up activities	–	–	International symposium jointly held with GelK of Kumamoto University and EDL of University of Tsukuba	International symposium jointly held with the Water Management Authority, GPSS-GLI of The University of Tokyo, GelK of Kumamoto University and EDL of University of Tsukuba
Required outcome	Unique proposals for local policymakers on solutions to water-related issues in the Heihe River basin			

Note: *GPSS-GLI* Graduate Program in Sustainability Science-Global Leadership Initiative; *GelK* International Joint Education Program for Groundwater Environmental Leaders; *EDL* Environmental Diplomatic Leader Program

Fig. 5.3 Integral organizational framework for the field exercise unit

for multiple issues, different methodologies from diverse fields are required. Although they cut across quadrants, natural science methods, including experiments and quantitative modeling, are mostly required to tackle the issues in the upper right quadrant. For the lower right quadrant, social science methods are mostly required. In the case of two left-side quadrants, humanity-based, hermeneutic methods are mostly required. Problem-based learning is the core concept of the course design. It brings up related issues and brings together the necessary research methodologies.

The competencies identified in Fig. 5.3 were not intentionally selected by us. They developed naturally in the process of participating in fieldwork, especially through group work. Team-based activities require the students to listen to, understand, and assimilate different ideas while contributing to groups from their respective fields and perspectives. Students need to find common research interests and decide on common research topics, as well as adapt to change, and finally to solve the problems. In addition, competencies spill over quadrants. For example, good communication skills may foster students' understanding of the varied concerns of the stakeholders, to create a shared vision, and to integrate methodologies and fields to find comprehensive solutions.

5.4 Experiences and Lessons Learned from the Oasis Unit

After four years, participants in the Oasis Unit generally consider that the unit has increased their understanding of the multiple dimensions of environmental issues and helped to improve their leadership skills/competencies. An et al. [23] reports the results of a questionnaire survey directed at Oasis Unit participants. The key points from this survey and students' comments after they joined the Oasis Unit are included as follows.

The questionnaire in An et al. [23] was adopted and modified from Gordon and Berry [24] to examine the educational effects of the Oasis Unit. The questionnaire was designed to monitor the students' way of thinking about environmental leadership, skill acquisition, pedagogy etc., before and after their participation of the field exercise. The questionnaire contains eight statements that reflect the most common current ideas about leadership. The respondents were asked to choose from five different points on a scale ranging from "strongly agree" to "strongly disagree." The results revealed obvious changes in awareness before and after the students participating in Oasis Unit. The Oasis Unit participants identified the participatory, open model and visible leadership through "field exercise." In addition, most of the Oasis Unit participants put more emphasis on the importance of leadership education development through practical experience. Oasis Unit participants tend to disagree with the statement that "leadership skills are inborn and intuitive." However, they also tend to have a higher awareness about the difficulties of developing environmental leaders. They seem to realize that participatory leadership requires time and resources which in turn requires a consensus among the stakeholders to achieve time-bound goals. This is presumably because in a field-oriented course, most of the students have their first experience of working with people from different cultural and academic backgrounds. In many cases, there is also a language barrier, which often hinders their ability to express their own ideas and to reach agreements with each other.

In general, former participants were satisfied with the content provided by the Oasis Unit. They think that a field-oriented environmental leadership course promotes positive cross-cultural interactions among students and faculty members. Many of them also mention that the involvement of local stakeholders and faculty members from different academic fields provides a comprehensive, balanced understanding of environmental issues. The students feel that their leadership skills were developed throughout the unit.

However, the Oasis Unit also has some clear shortcomings. The main problem is how to limit the amount of knowledge conveyed in advance by faculty members as well as a guarantee of the depth of the understanding of students. During the past four years, since we have put an emphasis on student initiatives and consensus building, the faculty members have been trying to limit the contents of pre-survey lectures to relevant fundamental knowledge. The students are required to find specific research questions and make detailed survey plans by themselves. Often, they start to realize after departure that their research proposals do not reflect local realities. The students sometimes need to make major adjustments to research proposals after departure. As a result, they often have to spend lots of time on discussions and the final survey appears to lack the depth and scope due to time constraints.

5.5 Concluding Remarks

The complexity of current environmental problems is triggering mounting concerns about the integration of academic fields to find solutions for sustainability of human future. There arises the need to equip students with a wide range of knowledge base and practical skills in terms of social, economic, cultural and physical dimensions of environmental issues. Undoubtedly, in the absence of established models, it is a challenge to move toward a holistic view for educational programs. In this chapter, we introduced a framework developed from Ken Wilber's four quadrant approach. This framework was built upon the core concept of problem-based learning. It has been adopted by one field-oriented course for environmental leadership development: the Oasis Unit, initiated by APIEL, The University of Tokyo. After four years of implementation, we have found that program participants were enthusiastic about and satisfied with the course. Therefore, we conclude that the use of the Integral Approach is effective for not only understanding complex environmental issues, but also the development and management of environmental leadership education programs. However, at the practical level, we also see difficulties with maintaining the depth/effectiveness of course content as well as a guarantee for students to use their initiative and to develop leadership skills within a limited time.

Acknowledgments A number of people have been part of the discussions for developing the field exercise unit conducted in the Heihe River basin. Especially highly acknowledged are the initiatives by Prof. Xin Li and Prof. Mingguo Ma (Cold and Arid Regions Environmental and Engineering Research Institute, Chinese Academy of Sciences), Prof. Guangwei Huang (Sophia University), Prof. Takashi Mino, Prof. Eiji Yamaji, and Prof. Tomochika Tokunaga (The University of Tokyo) for the start-up of the field exercise unit. However, the authors take sole responsibility for any errors and the interpretation provided in this chapter.

References

1. Akiyama T, Li J, Onuki M (2012) Integral leadership education for sustainable development. J Integr Theory Pract 7:72–86
2. Wilber K (2000) A theory of everything: an integral vision for business, politics. Science and spirituality. Shambhala Publications Inc., Boston
3. Wilber K (2007) Integral spirituality: a startling new role for religion in the modern and post-modern world. Shambhala Publications Inc., Boston

4. Esbjorn-Hargens S, Zimmerman ME (2009) Integral ecology: uniting multiple perspectives on the natural world. Shambhala Publications Inc., Boston
5. Eddy BG (2005) Integral geography: space, place, and perspective. World Futures 61:151–163
6. Kayane I, Miyazawa T, Zhu A (2006) Reikoukojou no mizu to shakai. Water Sci 291:41–72 (in Japanese)
7. Kayane I (2008) Environmental problems in modern china: issues and outlook. In: Kawai S (ed) New challenges and perspectives of modern Chinese studies. Universal Academy Press Inc., Tokyo, pp 265–285
8. Kayane I (2008) Reikoukojou to Tsuwano no hikaku kara kangaeta tourism no yakuwari. In: Database of international center for Chinese studies. Aichi University, Aichi (in Japanese)
9. Esbjorn-Hargens S (2005) Integral ecology: the what, who, and how of environmental phenomena. World Futures 61:5–49
10. Voros J (2001) Reframing environmental scanning: an integral approach. J Futures Stud 3:533–551
11. Floyd J, Zubevich K (2010) Linking foresight and sustainability: an integral approach. Futures 42:59–68
12. Akiyama T, Li J, Kubota J, Konagaya Y, Watanabe M (2012) Perspectives on sustainability assessment: an integral approach to historical changes in social systems and water environment in the Ili River Basin of Central Eurasia, 1900–2008. World Futures 68:595–627. doi:10.1080/02604027.2012.693852
13. Brown BC (2006) The use of an integral approach by UNDP's HIV/AIDS group as part of their global response to the HIV/AIDS epidemic. http://www.integralworld.net/pdf/Brown.pdf. Accessed 9 May 2010
14. Hochachka G (2008) Case studies in integral approaches in international development. J Integr Theory Pract 3:58–108
15. Gidleya JM, Hampson GP (2005) The evolution of futures in school education. Futures 37:255–271
16. Lloyd DG (2007) An integral approach to learning. Int J Environ Cult Econ Soc Sustain 2:27–33
17. Postel S (1999) Pillar of sand: can the irrigation miracle last? W. W. Norton & Company, New York
18. Shiklomanov IA (2000) Appraisal and assessment of world water resources. Water Int 25:11–32
19. Inoue M (2007) Shindaiyouzeinenkan ni okeru Kokuga no danryu to kokugakinsuiseido ni tsuite. In: Inoue M, Kato Y, Moriya K (eds) Oasis chiikishi ronsou. Shoukadoh, Japan (in Japanese). pp. 173–191
20. Inoue M (2010) Development of mountain forest and its environmental impact in the upper reaches of Heihe River basin of Ming and Qing period. J East Asian Cult Interact Stud 6:475–488 (in Japanese)
21. Wang X, Gao Q (2002) Sustainable development and management of water resources in the Hei River basin of North-west China. Water Resour Dev 18:335–352
22. Fang C, Bao C, Huang J (2007) Management implications to water resources constraint force on socio-economic system in rapid urbanization: a case study of the Hexi Corridor, NW China. Water Resour Manag 29:1613–1633
23. An KJ, Akiyama T, Kim JY, Hoshiko T, Furumai H (2011) The influence of field-oriented environmental education on leadership development. Procedia Soc Behav Sci 15:1271–1275
24. Gordon JC, Berry JK (2006) Environmental leadership equals essential leadership: redefining who leads and how. Yale University Press, New Haven
25. Akiyama T, Li J, Tokunaga T, Onuki M, An KJ, Hoshiko T, Ikeda I (2010) Integral approach to environmental leadership education: an exploration in the Heihe River basin, Northwestern China. In: Proceedings of the 8th international symposium on Southeast Asian water environment, pp 40–49

Chapter 6
Environmental Leadership Development Based on Activity Theory for Sustainable Urban Development in the Greater Pearl River Delta, China

Kyoungjin J. An

Abstract As cities in the GPRD develop, they go through an environmental transition associated with changes in the type of environmental challenge. With its growing economic power, China is playing an important role in the global economy, although the prosperity of the cities seems to have come with certain hurdles: social inequality and environmental deterioration. In response to such transaction, we have examined the GPRD's urban development, including urban formation, industry relocation, economic development, social inequality, and biodiversity conservation for a better quality of life (QOL). As a catalyst for sustainable urbanization, environmental leadership—where the urbanization process can be in harmony with the urgency for the economic development and sustainable future of the city—was explored. The holistic framework of activity theory was also applied within the area of sustainability to arrive at an inclusive structure for environmental leadership.

Keywords Environmental leadership development • Quality of life • Sustainable urban development

6.1 Introduction

The Greater Pearl River Delta (GPRD) is a megacity region in southern China, as shown in Fig. 6.1. It includes nine municipalities, which are major component cities of the GPRD in Guangdong Province, and two special administrative regions: Hong Kong and Macao. Currently, the GPRD has a population of over 50 million and

K.J. An (✉)
Asian Program for Incubation of Environmental Leaders (APIEL),
Department of Urban Engineering, Graduate School of Engineering,
The University of Tokyo, 7-3-1 Hongo, Bunkyo-ku, Tokyo 113-8656, Japan
e-mail: ankj@env.t.u-tokyo.ac.jp

T. Mino and K. Hanaki (eds.), *Environmental Leadership Capacity Building in Higher Education*, DOI 10.1007/978-4-431-54340-4_6,
© The Author(s) 2013

Fig. 6.1 Megacity region of China's Greater Pearl River Delta (GPRD) (*Source*: Invest HK, www. invest.gov.hk)

covers a land area of around 43,000 km². About 20 years ago, the GPRD portion in Guangdong Province largely contained underdeveloped rural villages. Since then, there has been an enormous transformation in the GPRD. The "reform and opening up" policy of China has had a dramatic impact on society and the economy over the past 30 years, allowing the country to enter an era of rapid development. Megacity regions, like the GPRD, are one of forerunners in this transformation.

But this rapid change has led to a development that is not only skewed towards economic improvement but it has also created a society with disparities and deprivation. Economically, the GPRD has developed into a renowned world factory, while Hong Kong has been restructured as a regional service center, providing industries in the GPRD with frontend functions, such as research, marketing, and distribution. A regional division of labor, the "front shops, back factories" model began to take shape in the 1990s. Spatially, the GPRD as a whole has become increasingly polycentric; many cities and towns that were formerly peripheral and rural areas have developed into active economic centers. The polycentric spatial form has combined with the rise of urban entrepreneurialism, resulting in a rapidly developing political environment that encourages cities to compete against one another for mobile capital.

The undesirable consequences of political fragmentation are becoming more and more acute. The impacts of Hong Kong and Macao, under the "one country, two systems" model, are added complications to this fragmentation. Political borders stand in the way of coordinated planning. Socially, the GPRD has faced challenges caused by an increase in the resident population, which has overwhelmed the GPRD governments. The flood of rural to urban migration has weighed down the infrastructure in cities and has led to tremendous growth of un-serviced urban areas where millions of migrant workers lack access to basic services. There is also a pressing need to address the problems of widespread misuse of land, urban sprawl, traffic congestion, poor sanitation, and the declining living environment in all cities, especially those that are threatened by rapid and often uncontrolled growth, inadequate and poorly maintained infrastructure, industrialization, and the increasing ownership of cars and motorcycles.

Sustainability in urbanization is closely linked with competitiveness especially economic competitiveness [1]. However, competitiveness does not include non-economic success or accept the consequences, such as social polarization and environmental pollution [2]; favoring economic growth has wider problematic social consequences [3]. Therefore, the concept of competitiveness is being modified to incorporate social and environmental criteria as it directly affects the quality of life. For example, a study by Jiang and Shen [1] suggests that Guangzhou's overall competitiveness among 20 Chinese cities is falling due to the lower social and environmental performance. Perhaps, Guangzhou's competitiveness demands a balance of economic growth with social and environmental performance, which in turn significantly affects the quality of life.

The Asian Program for Incubation of Environmental Leaders (APIEL) of The University of Tokyo (UT) has set out to help understand this balance. As it was introduced in Chap. 1, APIEL is an educational program designed to foster environmental leaders, especially aimed at sustainability issues in Asia. This chapter describes the framework and discusses the circumstances under which activity theory could be used for an environmental leadership program and to help build a better QOL in the cities, within the rapidly urbanizing GPRD region. This field unit took up environmental leadership as a tool for improving the QOL during rapid urbanization as well as sustainable development in the GPRD. In addition, this field unit illustrates the use of environmental leadership for a sustainable future in GPRD cities, using the activity theory framework. Case studies, mostly focusing on major domains for urban QOL, explore the intricate relationship with urbanization. Methods based on activity theory were used to conduct the field unit and manage the data collection and analysis processes. This chapter concludes with a discussion of the relevance and suitability of activity theory as a framework for the current complex problems in fostering future environmental leaders.

6.2 Activity Theory Framework for Building Leadership Capacity

Activity theory is a socio-psychological theory with roots in the work of the Russian psychologist Lev Vygotsky during the first half of the twentieth century. Vygotsky's [4] important insight into the dynamics of consciousness was that it is essentially subjective and shaped by each person's social and cultural experience. In addition, Vygotsky [4] saw human activity as distinct from non-human entities; it is mediated by tools, the most significant of which is language. Vygotsky's [4] work was continued by others, among them Leont'ey [5], who developed a conceptual framework for a complete theory of human activity. According to Leont'ey [5], activity is a system that has structure, its own internal transitions and transformations, and its own development.

Essentially, Vygotsky [4] defined human activity as a relationship between subject and object, i.e. a person working at something. In this dynamic, purposeful relationship the "always active" subject learns and grows, while the object is interpreted and reinterpreted by the subject in the ongoing conduct of the activity. Approach of activity theory is promising for environmental leadership because it arrives at logical conclusions through a complex interpersonal process of transfer knowledge from individual to group. Engeström [6] gave a more concrete expression to this structure in the triangular representation, which is commonly used to depict an activity. The core of an activity is the relationship between subject (human) and object (purpose) mediated by tools and community. This is a two-way concept of mediation where the capability and availability of tools mediates what is able to be done; the tools, in turn, evolve to hold the historical knowledge of how the community behaves and is organized. The third-generation of activity theory, proposed by Engeström [7] in 2001, advanced the idea of internal contradictions as the driving force for change, diversity and dialogue from divergent perspectives within an activity system, and of networks of interactive activity systems.

Activity theory today is based on the idea that people change or learn when they engage in productive activity, but in productive activity they also change their system. For fostering future leaders, activity theory suggests that leadership occurs through the interaction of the leader with other components of an activity system, such as the tools the leaders have available and the people with whom they interact in a division of labor. Thus, a leader is directed toward a particular goal or outcome. Krasny and Roth [8] explained that any one activity system may overlap with another system, for example, a school and a watershed organization both focus on taking measurements with the goals of producing a report for a class and impacting environmental policy, respectively.

6.3 GPRD Implementation

The search for a sustainable future has made fostering environmental leaders essential, especially for mediating between knowledge societies and the community. Studies have shown three behaviors that appear relevant for environmental leadership: articulating an appealing vision with environmental elements, changing perceptions about environmental issues, and taking symbolic action to demonstrate a personal commitment to environmental issues [9]. During the GPRD field exercise, APIEL also emphasizes six attributes for environmental leadership education: *information*, or the ability to find, understand, and transmit needed information;

inclusion, or listening and using all available skills and ideas; *decision*, or defining and pursuing an action agenda; *dispatch* (action), or doing things now rather than later; *standard setting*, or formulating the definition of success; and *humanity*, or using empathy and humor in dealing with others [10].

This chapter will mainly discuss how the students identified a problem and selected a study area for group work during the field exercise provided to practice and build their leadership. The final outcome presented "Sustainable urban development toward Green City: the Greater Pearl River Delta (GPRD), China." in Year 2012 was chosen to describe the challenges during the fieldwork and to reflect the feedback.

6.3.1 Program Outline

Over the last 4 years, APIEL has built broad, resonant networks among universities in Asia. The ties with the Hong Kong University of Science and Technology (HKUST) and Sun Yat-sen University (SYSU) grew especially strong by conducting a field exercise called the GRPD Unit over three consecutive years. During the first year of the program (2010), using case studies, we covered environmental issues and the need for environmental leaders in Asia. Broad discussion topics included the environment, the need for environmental leaders, leadership examples in Asia, as well as a sustainable environment and management in the GPRD.

Then, over the next 2 years, the program focused on sustainable urban development in the GPRD. To start, sustainable urban relocation and regeneration of industrial regions in the GPRD were discussed. Since the GPRD is one of the leading economic regions in southern China and a major manufacturing center—combined with the booming economy and Western influences from Hong Kong—it has created an economic gateway attracting foreign capital into mainland China. In this regard, several topics were chosen for students to discuss, and in doing so, build their leadership skills. These topics included trans-boundary issues and collaborative programs to tackle regional air pollution (in turn, to cope with climate change), urban regeneration, and industry relocation for sustainable development in the GPRD.

In the third year, we reviewed the approaches that have been used for the GPRD's urban development, and related environmental loading over the last three decades. An overview of the approaches used in guiding the GPRD's urban development were studied and discussed. In addition, related environmental loading for the past three decades was studied through a lecture and a field trip during the unit. In March 2008, Sir Donald Tsang Yam-kuen, the chief executive of Hong Kong, proposed to the Guangdong party secretary, Wang Yang, that the two territories should jointly form a "Quality Living Circle in the Green Greater Pearl River Delta." The guiding principles would be promoting environmental protection and sustainable development. Table 6.1 summarizes the outline of the field exercise conducted in Year 2012. Students examined GPRD's urban development, including urban formation, industry relocation, economic development, social equity, and biodiversity conservation.

Table 6.1 Outline of the GPRD Unit field exercise (2012)

Title	Sustainable urban development toward Green City: the Greater Pearl River Delta (GPRD), China
Keywords	Environmental leadership development, sustainable urban development, Geographic Information System (GIS), quality of life (QOL)
Date	16–25 February 2012
Place	Hong Kong and Guangzhou, China
Participants	24 students from 8 countries—APIEL: 6 students, 3 from the Department of Urban Engineering (UE) and 3 from the Graduate Program in Sustainability Science (GPSS); Hong Kong University of Science and Technology (HKUST): 5 students from engineering; Sun Yat-sen University (SYSU): 7 from science; Seoul National University (SNU): 6 from business, international relations and urban planning
Collaborating partners/ organizations	(1) HKUST, SYSU, Asian Institute for Energy, Environment and Sustainability (AIEES) at SNU, and Hong Kong Green Building Council
Information lecture	(1) Urbanization and migration in China; (2) moving to low carbon society in cities; (3) designing high density cities for sustainable and quality living—a few notes for architects, planners and policy makers; (4) environmental GIS applications; (5) global and local environmental loading and quality of life; (6) urban development and sprawl; (7) environmental leadership and GPRD activity theory; (8) green city planning at the city level in the GPRD; (9) city of Guangzhou Urban Land Administration Bureau; (10) rapid urbanization, urban forms and energy consumption in the Pearl River delta; (11) GPRD urban development and its environmental loading; (12) understanding the urban design approach in the process of city planning
Fieldwork	(1) Hong Kong Science Park at Shatin; (2) Hong Kong-Shenzhen-Dongguan City Exhibition Hall-Guanzhou SYSU; (3) change of GPRD urban form: Nansha wetland, Greenway in Guangzhou; (4) change of GPRD urban form: Donghaochong, Liwanchong, Shangxiajia, Haizhu Lake; (5) case study area visit and interview/GIS practice
Group work	(1) Leadership attributes: communication; (2) GPRD Strengths–Weaknesses–Opportunities–Threats (SWOT) analysis, QOL and GIS; (3) develop the indicators for survey questionnaire for quality of life in GPRD; (4) vision and action learning; (5) preparation of final presentation
Case study area in Guangzhou	Industry zone (Honda factory), a higher education mega centre (HEMC, Xiaoguwei island), newly developed town (Zhujiang), well preserved old town (Yangzhong Street village) and an urban village (Xiadu village)
Conference presentation	The First International Conference for International Society of Habitat Engineering and Design (ISHED), Asia City in the New Age: (1) Urban Forms and Pro-Environmental Behavior for Waste Management: a Case Study in Guangzhou, China, [11] and (2) Rapid Transformation and Change in Local People's Quality of Life: A Case Study of Xiaoguwei Island, Guangzhou, China [12]

6.3.2 Applying Activity Theory to GPRD Field Exercise

As learning activities are dynamic and may have multiple connections among
their own elements and with other activity systems, we carefully applied activity
theory to design the APIEL education program for urban development in the

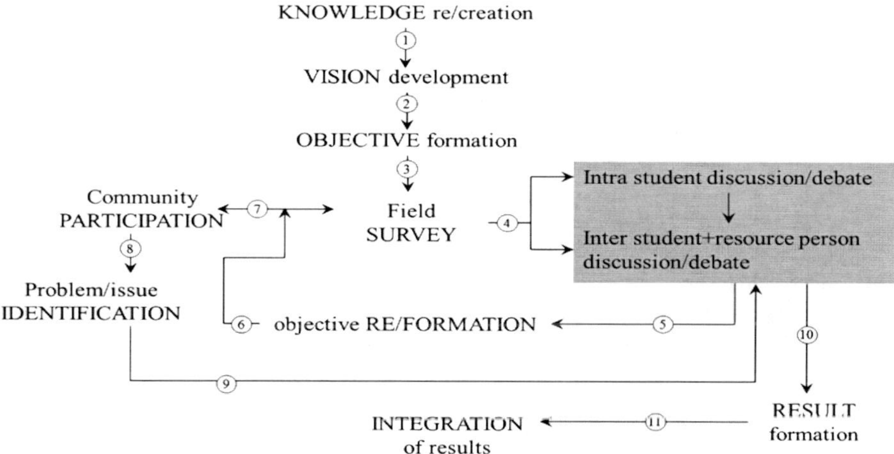

Fig. 6.2 Chronological metrics methodology for leadership development

GPRD. This approach also aided the process of fostering change agents and incubating environmental leadership skills in students. A chronological metrics methodology was developed and is shown in Fig. 6.2. An explicit environmental issue prevalent in Asian cities is used in this leadership education approach. Then, the resource people work with graduate students from multiple disciplines to make explicit the theory and assumptions implicit in the issues they are undertaking. An area is chosen for case studies where the issue is most pressing. After faculty members and students arrive at a consensus about the potential problems and issues, a field unit to the site takes place. A realistic approach is combined with a problem in pre-field unit sessions. Put another way, using the activity theory of change approach, the faculty members seeks evidence that the assumed (or theorized) links between program activities or processes and the desired results can be borne out by the field trips and experience. Faculty members then compare the program theory about how an intervention will unfold with the observations that they make about how it actually did unfold [13].

In February 2012, the APIEL GPRD field exercise undertook field-oriented work on the challenges of QOL in the rapidly urbanizing GPRD region of China. Environmental leadership using an activity theory approach was used for studying the sustainable future of the region. In particular, Guangzhou, where rapid urbanization is taking place, was chosen as the "object" intended to expose students to the scope of problem solving in a city facing severe environmental issues associated with social and economic dimensions. Activity theory and "resonance" were integrated to arrive at final outcomes, as shown in Fig. 6.3.

Students from multiple disciplines and researchers from four premier institutions in Asia—UT, HKUST, SYSU and SNU—playing the role of "subject," came together at Guangzhou to study a sustainable future for the region. Environmental leadership based on ethics, information, and knowledge was extended with tools, such as surveys, geographic information systems (GIS), and divisions of labor

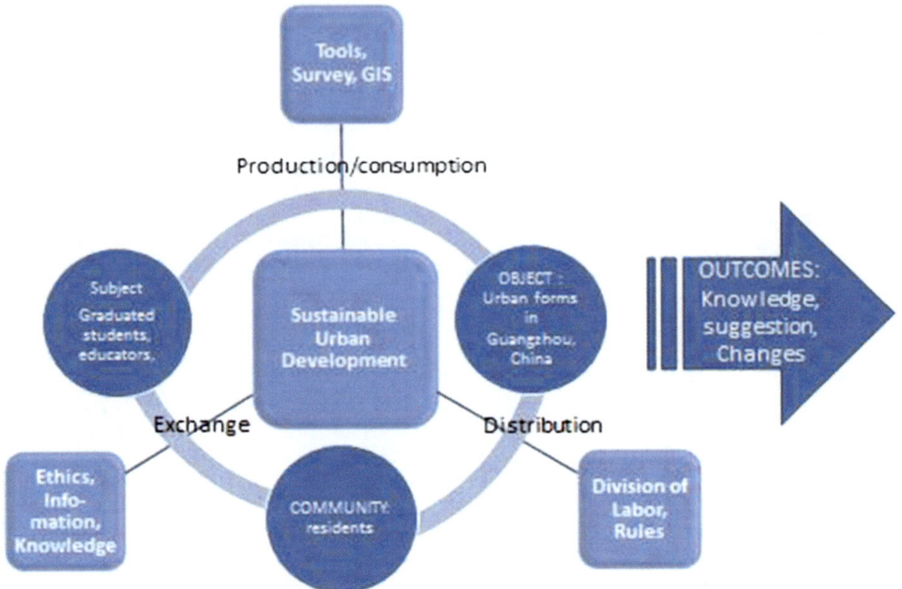

Fig. 6.3 Framework of leadership development for sustainable urban development in Guangzhou

within the program. The final outcome was presented as a report, presentation, and a journal paper.

6.3.3 The Selection of a Study Area

Guangzhou is in southern China on the Pearl River Delta plain. The city of Guangzhou has ten districts (Yuexiu, Dongshan, Haizhu, Liwan, Tianhe, Baiyun Whampoa, Fangchun, Huadu and Panyu) with an area of approximately 3,718.8 km² (about 50% of the total municipal area) with a census population of 7.4 million (74% of the total municipal population) [14]. Land use change from 2003 to 2008 in Guangzhou was analyzed using GIS to investigate how urbanization was taking place over a decade.

Dramatic changes in the built-up area over 5 years were compared with other land use, such as water bodies, arable land, transportation, and forests, as illustrated in Fig. 6.4. The built-up area in 2003 is in orange (agglomerated in central Guangzhou); the area in black shows not only the old town but also all over the city as a new town in 2008. It was very clear that the city underwent significant change in land use between 2003 and 2008 due to rapid urbanization. Among the major changes, the most prominent was the increase in built-up area (about 80% compared with data from 2003). On the other hand, as shown in Fig. 6.5, water bodies and arable land were not conserved but were taken over by built-up areas.

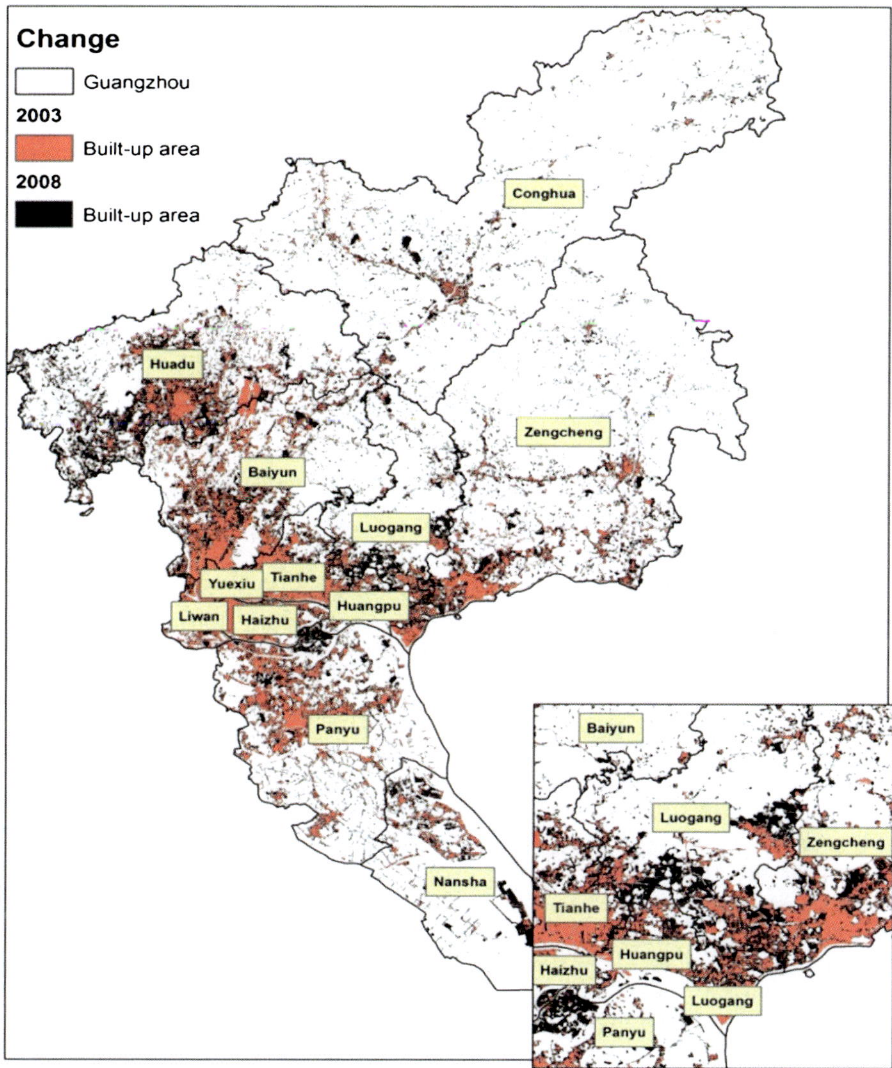

Fig. 6.4 Increase in built-up area Guangzhou from 2003 to 2008

Mindful of this urban development, we conducted empirical studies in the Guangzhou to investigate QOL using survey-based approaches to assess a sustainable method for urban development. The survey, including objective and subjective data, was integrated with a GIS tool to enhance the investigation of QOL. Students selected a study area and did group work on site. To begin with, the city of Guangzhou, as a study field, was divided into five areas according to their urban form: industry zone (Honda factory), a higher education mega centre (Xiaoguwei island), newly developed town (Zhujiang), well-preserved old town (Yangzhong

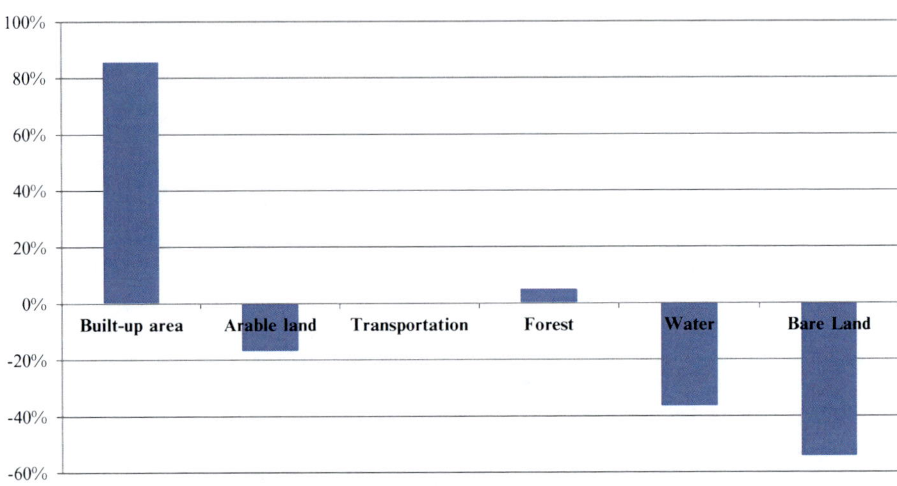

Fig. 6.5 Land use change (%) in Guangzhou from 2003 to 2008

Street village), and an urban village (Xiadu village) for newcomers seeking economic opportunities, as shown in Fig. 6.6.

The selection of a study area for each group was based on the vision created by the group members through communication, knowledge creation, and objectives. This process was challenging because students from different backgrounds, countries, major subjects of study, and personal characteristics had to build trust through exchange and then divide up roles, as illustrated in Fig. 6.3. Through the lens of QOL, each urban form for Guangzhou's urban development was measured and extended to the pro-environmental behavior of the residents. Learning occurred through the interaction of the learner with the components of social sustainability in the field. The creation of a vision, division of labor, and generating a knowledge base through reconnaissance and sample surveys, were used for arriving at particular policy-level outcomes.

6.3.4   Creating Vision: Challenges from Student's Feedback

This section was abstracted from student reports expressing the challenges they faced during the unit. First, most students said that arriving at a consensus for a common goal was a challenge. A student wrote in her final report that the group's decision on the vision was a spontaneous process of continuous discussion, and that a systematic division of roles was the next step in the process. In her group, vision creation was done through personal interest, then a tree of functions was drawn and every member of the team was asked to write their names beside their choice for work. This approach was then later adopted as a common methodology. Post-analysis and brainstorming were used in her group to focus on waste management

Fig. 6.6 Urban forms in Guangzhou [11]

and the pro-environmental sector. Another student also emphasized this challenge, saying that "after the first reconnaissance survey, we understood that the major challenge which we faced was to narrow down the study objectives, as we all found issues were myriad." After deciding on the focus, sketching out the framework and choosing a methodology for the study were found to be easy and his group proceeded smoothly. Students addressed important findings through activities, saying, "From the beginning we tried to remain focused on the objectives that we were trying to tackle. At times, we did deviate from our main point of view, but with guidance from resource persons, we got back on track. Importance of multi-disciplinary approach was such an essential tool; we wouldn't have realized that before." Whenever his group understood the gaps that each one of them had, they then tried to identify the potential among the team. Then all members tried to maximize the quality of output through patient listening, discussing, debating, and taking notes. To begin with, members differed on the shared vision. Students also mentioned that each of the members in her group had a different focus of concern at the beginning of the program, so they spent quite a long time on group discussion but could not arrive at a common vision. However, more knowledge about Guangzhou during the lectures and fieldwork helped them make a breakthrough in vision creating. All groups had members from different countries and academic backgrounds.

Communication and mutual understanding were found essential to work with people from diverse backgrounds and to arrive at a common vision. Eventually, the students came together on the object of the study and worked together, despite. Sustainable urban development as a catalyst of leadership capacity building in this program involves many factors, including the environment, economics, and society. At the same time, many stakeholders' viewpoints have to be considered.

6.3.5 Challenges from Educators for Appraisal

The success of this program was evaluated in two ways: the report and/or paper students produced based on the outcome of the program and a self-evaluation of leadership. As this program emphasizes not only an environmental problem solving but also the process of handling and solving them, and in so doing develop leadership, it is challenging to measure how much the students developed their leadership skills. Faculty members have asked students at the end of program: How did you create a shared vision with others? What was the problem(s) you and your group members identified? What was the strategy you built up to solve them? Did you create a vision table, information needs diagram, group matrix, action agenda, and implementation plan? Did those lead you to an integrated problem solution? Did you commit to being a standard setter? Did you positively study and practice essential leadership attributes? Did you find the areas of your strength through participation in this program? Did you find the areas of your weakness through participation in this program? and How are you going to apply your leadership skills when you are an environmental leader in the future? Table 6.2 shows the self-evaluations of leadership after the unit from six UT

Table 6.2 Self-evaluation of leadership after the unit from UT students (six students)

		Standard met	Standard met with concerns	Standard not met
(1)	Creating a clear vision to solve an environmental problem	5		1
(2)	Applying information to solve the problem	5	1	
(3)	Developing communicating, listening, and interpersonal skills	6		
(4)	Use of resources and facilities	4	2	
(5)	Decision: ability to define and pursue an action agenda	5	1	
(6)	Action: ability to take on tasks NOW, rather than later	5	1	
(7)	Collaborating with others	6		
(8)	Sense of humor during the unit	5	1	
(9)	Understanding lectures and fieldwork	4	2	
(10)	Overall achievement	5	1	

students. Though we found that one student couldn't meet the standard for creating a clear vision to solve an environmental problem, communication and collaboration standards were met by all students. This implies that among the leadership attributes that we have taught, inclusion was found important in our program.

For the academic achievement, the two presentations listed below were made at international conferences:

Md. Manjur Morshed, Kyoungjin J. An, Xu Xiaohan, Hironori Hamasaki, Tomomi Hoshiko (2012) Rapid Transformation and Change in Local People's Quality of Life: A Case Study of Xiaoguwei Island, Guangzhou, China. The 1st International Conference on Habitat Engineering and Design, October 13–14, 2012, Tongji University, China. Oral Presentation.

Jungchan Lee, Kyoungjin An (2012) Urban Forms and Pro-Environmental Behavior for Waste Management: a Case Study in Guangzhou of China, The 1st International Conference on Habitat Engineering and Design, October 13–14, 2012, Tongji University, China. Oral Presentation.

6.4 Conclusions

Environmental leadership in this context is about personal growth or change within a group to guide positive development toward a vision of an environmentally friendly and better future. Our belief, based on many years of professional experience, is that leadership can be learned. Indeed, we think that all participants who joined the GPRD Unit will lead others at some time and place in their career, and that overt preparation for leadership should be an integral part of professional education and experience. Many traditional leaders are given high marks for accomplishing much in a short time, and moreover, have been first and foremost, effective adversaries of competing causes; slaying dragons has always been a sure route to leadership success. However, in this chapter we argue that environmental problems are not most effectively solved through processes. Finding common ground, negotiation, and cooperation are better suited to most of the complex, long-term problems facing environmental leaders.

This chapter examined how educators, students, and other people in higher academic settings can work together to build transformational leadership capacity to help drive more sustainable urban development practices. In relation to university-based higher education, we linked customized leadership development activities (based on activity theory) with prominent urbanization issues in Guangzhou. This provided a system to produce educational reports, papers, and conferences. The process of building knowledge and skills as well as social networks, then finally led to behavioral change in a variety of proposals presented by students. Thus, this chapter argues that building transformational leadership capacity within a higher academic setting is another way educators can lead our society to desirable behavioral change and sustainable urban development. Such leadership development initiatives are essential in contemporary environmental learning, and they can be both meaningful and worthwhile.

This indicates that the APIEL can go alongside a socially critical approach, but more importantly, it focuses on experience where fieldwork takes place within an activity system. We have observed how students assimilate an approach that reflects the complexity of thinking and learning in relation to the environment. Feedback from multiple sources was collected and used to evaluate and improve the activity. This research is both ongoing and challenging.

Acknowledgments The GPRD Unit was successfully conducted with solid support from Sun Yat-sen University and Hong Kong University of Science and Technology, and special thanks are extended to Prof. Shiyu Li at SYSU and Prof. Guanghao Chen at HKUST for their strong support.

References

1. Jiang Y, Shen J (2010) Measuring the urban competitiveness of Chinese cities in 2000. Cities 27:307–314
2. Lever WF, Turok I (1999) Competitive cities: introduction to the review. Urban Stud 36:791–794
3. Wang J (2009) Art in capital: shaping distinctiveness in a culture-led urban regeneration project in Red Town, Shanghai. Cities 26:245–254
4. Vygotsky LS (1978) Mind in Society: the development of higher psychological processes. Harvard University Press, Cambridge

5. Leont'ey AN (1981) Problem of the development of the mind. Progress, Moscow
6. Engeström Y (1996) Developmental work research as educational research, Nordisk Pedagogik. J Nordic Educ Res 16(5):131–143
7. Engeström Y (2001) Expansive learning at work: toward an activity theoretical reconceptualization. J Educ Work 14(1):133–156
8. Krasny ME, Roth WM (2010) Environmental education for social-ecological system resilience: a perspective from activity theory. Environ Educ Res 16:545–558
9. Portugal E, Yuki G (1994) Perspectives on environmental leadership. Leadersh Q 5:271–276
10. Gordon JC, Berry KJ (2006) Environmental leadership equals essential leadership. Yale University Press, New Haven
11. Lee JC, An KJ (2012) Urban forms and pro-environmental behavior for waste management: a case study in Guangzhou of China. In: The 1st international conference on habitat engineering and design, Oral presentation, Tongji University, China, 13–14 October 2012
12. Morshed MM, An KJ, Xu XH, Hamasaki H, Hoshiko T (2012) Rapid transformation and change in loc.l people's quality of life: a case study of Xiaoguwei Island, Guangzhou, China. In: The 1st international conference on habitat engineering and design, Oral presentation, Tongji University, China, 13–14 October 2012
13. Weitzman BC, Silver D, Dillman KN (2002) Integrating a comparison group design into a theory of change evaluation: the case of the urban health initiative. Am J Eval 23(4):371–385
14. Xu J, Yeh AGO (2003) City profile: Guangzhou. Cities 20:361–374

Chapter 7
Environmental Leadership Development:
A Cambodian Case

Hironori Hamasaki and Hiroyuki Katayama

Abstract This chapter presents an overview of the activities of Asian Program for Incubation of Environmental Leaders (APIEL)'s Cambodia Unit and its self-evaluation. First, we clarify the reason why Cambodia was chosen as a location to implement APIEL's field exercise unit. Second, the basic concept and approach are introduced. Third, program components that include preliminary tasks, on-site program, and subsequent presentations and reports by students are shown. Finally, self-examination is given to propose recommendations to improve the Cambodia Unit.

Keywords ODA • Solid waste management • Sustainable development • Tourism • Water supply

7.1 Introduction

Cambodia's economic growth has been expected to accelerate to its fastest pace over a decade by a burst urban construction funded by foreign aid and investment. According to the International Monetary Fund, Cambodia's yearly average GDP growth rate over the last 10 years is 7.7% [1] and it will bring new investments. Likewise many other developing Asian countries where the economy is growing rapidly, environmental

H. Hamasaki
Research Institute for Humanity and Nature,
Kyoto, Japan
e-mail: hamasaki@chikyu.ac.jp

H. Katayama (✉)
Department of Urban Engineering, Graduate School of Engineering,
7-3-1 Hongo, Bunkyo-ku, Tokyo 113-8656, Japan
e-mail: katayama@env.t.u-tokyo.ac.jp

T. Mino and K. Hanaki (eds.), *Environmental Leadership Capacity
Building in Higher Education*, DOI 10.1007/978-4-431-54340-4_7,

problems, such as water contamination, air pollution, and chaotic urbanization, have risen to the surface and are affecting people's lives and health in Cambodia. In addition, this country lacks both of physical infrastructure and human resource. It is urgent to draw the nation's pathway to build a sustainable society for the next generation.

Cambodia, on the other hand, has its unique history symbolized by Angkor Wat Temple and the Pol Pot regime. The Angkor Wat Temple, one of the most conquering kings of the Khmer Empire, Suryavarman II, constructed in the beginning of the twelfth century, is considered as one of the largest temples in the world. The number of visitors has been increasing substantially in recent years, and the provincial tourism department reported in the year 2012 that about two million foreign tourists visited this World Heritage site. However, the municipal government of Siem Reap that holds jurisdiction over Angkor Wat has been suffering from problems such as water supply, solid waste management, heavy traffic, etc. On the other hand, the Pol Pot regime was built by a communist party, Khmer Rouge, in 1970s. They attempted to build a farming society. However, it resulted in the death of 25 % of the country's population including those highly educated from starvation, overwork and executions (genocide). Even today, some are afraid of the re-emergence of a similar political reign. This sense of fear among people is still causing serious social issues such as low school enrollment ratio, and delaying the sound development of Cambodia, leaving the country in tough environment in every aspect.

These facts provide a hint why we chose Cambodia for one of APIEL's field exercise units' study site: this country was considered to be one of the best places where students acquire a holistic view over environmental issues with social and economic insights.

7.2 Basic Concept and Approach of the Cambodia Unit

The theme of Unit "sustainable development in Cambodia" reflexes the complex nature of Cambodia's environmental issues. The United Nations says in the Brundtland Reports that "sustainable development is development that meets the needs of the present without compromising the ability of future generations to meet their own needs [2]." However, sustainable development in the context of Cambodia is wide-ranging and elusive.

Each student started with answering the core question that was "what is most important for sustainable development in Cambodia?" They had difficulty in the first place to respond this theme and struggled. However, eventually, they identified and addressed the emergent environmental issues to be solved through communication among group members. This process was what we intended. Environmental problems that were causing conflicts among stakeholders often had complicated relationships with other issues, such as poverty or social justice. The authors believe that the students should not choose their own topic of study for setting an easy answer. Students who wish to be environmental leaders must struggle throughout the discourse of problem identification with other stakeholders before finding *the* answer.

The main objective of the Cambodia Unit was to foster responsible environmental leaders who can bravely criticize and seek for fundamental solutions to the complex and broad environmental issues in Cambodia. In the course to find or define a problem, the students developed a very important quality: a sense of *responsibility*. The authors often encouraged students to enjoy struggling during the Unit. Why? In the real world, leaders have to go through a time consuming process of finding problems and solutions in order to bring positive changes to a society. Thus, this struggle at this phase was intended to encourage the students to improve their leadership qualities, namely the discussion and communication skills.

Authors observed how their opinions and ways of thinking would change during the implementation of this Unit. The result was quite interesting. At the beginning of the Unit, their opinions were based mainly on their own academic backgrounds and majors. In other words, they looked at problems from only one direction. However, the students' opinions after their trip became more multifaceted, and their reports were written from holistic viewpoints.

7.3 Contents of the Cambodia Unit

The Cambodia Unit consists of three parts: (1) preliminary lectures and assignments before visiting Cambodia, (2) fieldwork during a stay in Cambodia, and (3) final presentations and reports after the fieldwork. The details on each part are described in the following sections.

The basic idea of the Cambodia Unit is to combine one lecture with one associated site visit on the same day. One day, for example, we took up the case of Japan International Cooperation Agency (JICA), in the morning we had a lecture on JICA's operation in Cambodia, and then in the afternoon, we had a site visit that supplemented the details. Students seemed to recognize the importance of visiting and seeing sites. In other words, lectures provided rich information, but students were able to find more useful information by looking at real-life situations. Also, students uncovered the differences among themselves after visiting the same place, which brought out a good chance to experience the diversified ideas. This program construction reflected other intentions of Unit organizers. Students were supposed to understand the real situation of environmental issues in Cambodia as well as to think why those environmental problems are difficult to solve through site visits. This approach motivated students to investigate how big the difference between learning knowledge from textbooks and seeing things as they are with their own eyes is.

7.3.1 *Preliminary Lectures and Assignments*

Preliminary activities before the on-site program in Cambodia were significant for two reasons. First, preliminary lectures were necessary for students because they have never visited Cambodia or known that much about Cambodia. Thus, these lectures focused on providing fundamental information and knowledge such as historical, economic, social, and environmental issues. Above all, learning the history of the Pol Pot regime was indispensable for understanding Cambodia; Cambodian society even today is suffering from its aftermath. In addition to that, as a collateral benefit, participants from different universities were able to get to know each other through a teleconferencing system before meeting together in Cambodia. Second, preliminary assignments were helpful for students to start communication, discussion, and interaction with each other from an early stage of Unit implementation. Sharing academic background sheet of each participant was also useful for students to have better idea about the others, helping them to discuss and exchange opinions more smoothly, even when using Web- and e-mail-based communications.

The significance of the preliminary meeting was typically seen when we compare the implementation of the Unit in 2011 with that in 2012. In 2011, students from the Royal University of Phnom Penh (RUPP) could not attend preliminary lectures because they did not have access to a teleconferencing system. That year, it took some certain time for RUPP students to mix comfortably with others from The University of Tokyo (UT) and Seoul National University (SNU). In 2012, the RUPP students attended the preliminary lecture through a teleconferencing system, thanks to the cooperation of World Bank and JICA. This three-way meeting enabled students from three different universities to communicate smoothly from the beginning to the end of the field exercises in Cambodia.

7.3.2 Fieldwork in Cambodia

7.3.2.1 Lectures

The topics of lectures held in Cambodia were deliberately wide-ranged since the main theme of this Unit was sustainable development. We selected several rather general topics for lectures, including water supply, solid waste management, rural development, Official Development Assistance (ODA), tourism, etc. Naturally, stakeholders related to these lectures have diverse views.

In 2011, we had lectures from JICA, the Korean International Cooperation Agency, Community Sanitation and Recycling Organization, urban planning consultants, local government officials from the city of Siem Reap, and the Authority for the Protection and Management of Angkor and the Region of Siem Reap which is an organization managing Angkor Wat. Based on these lectures, all the groups discussed and decided on their own topics for their final presentations.

When we started the preparation of Cambodia Unit in 2012, we reviewed the program held in 2011 and found it necessary to prepare a lecture related to leadership. Therefore, we requested Mr. Ek Sonn Chan through local JICA office to give us a lecture (Fig. 7.1). He was appointed in 1993 as the General Director of Phnom

Fig. 7.1 Lecture on leadership by Mr. Ek Sonn Chan

Penh Water Supply Authority where he exercised strong leadership, and contributed to the reconstruction of the water supply system in Phnom Penh. In particular, he succeeded in collecting water fees from over 90 % of households, which is seldom the case with developing countries that suffer from weak or lack of governance. He is respected by the people in Cambodia and has been called an incredible person [3, 4]. His lecture gave some valuable insights on leadership to the students based on his experiences of struggle to overcome many obstacles. His message greatly impressed the students and helped them to appreciate the various aspects of leadership and their own lives.

7.3.2.2 Site Visits

The sites to visit were chosen carefully, taking the advices from lecturers. For example, for water supply and urban flood control projects arranged by JICA, we visited a water treatment plant as well as a pumping station, and observed the drainage system (Fig. 7.2).

In the Cambodia Unit, along with the specific environmental issues, students were required to think about the nation's history with a holistic viewpoint because of the impact of Pol Pot regime on various aspects of people's life in Cambodia.

Fig. 7.2 On-site lecture on flood control in Phnom Penh

Thus, the faculty members let students visit symbolic places of genocide, such as the "killing fields," Tuol Sleng Museum in Phnom Penh, and the "killing cave" in Battambang, that turned out to be an experience that was beyond description. In addition, the visit to Angkor Wat revealed another issue that Cambodia is facing the conflict between economic development through tourism and environmental preservation.

7.3.2.3 Group Work

We allocated time for group work almost every day for around 10 days to prepare for final presentations. That means, students have to communicate with each other every day (Fig. 7.3). They seemed surprised to have found that, even though they visited the same place, they had different ideas about what they saw, and were interested in different aspects of sustainable development in Cambodia. They also realized the importance of listening to and respecting the others through their group discussions. The students then had to make decisions, as well as organize their thoughts, for the final presentations.

As mentioned in Sect. 7.2, we let each group determine their own topic by themselves from wide-ranging topics related to sustainable development. We believe that this was a good training for them to cultivate leadership. It took almost all the

Fig. 7.3 Interview with JICA staff as part of group work

groups a lot of time to agree on their own topic. They "struggled," but later said in their feedback that this was a very good experience.

7.3.3 Final Presentations and Reports

It was a requirement for UT students to make final presentations and submit reports after the completion of on-site program. They made the presentations more informative and better structured than the ones they gave in Cambodia on the last day of the fieldwork. They re-arranged the contents and added some new materials they searched for after their return.

The final reports that the UT students submitted were interesting. The question for the final report was "What is most important for sustainable development in Cambodia?" the same question we asked for the preliminary assignment. Their answers were based mainly on their own academic backgrounds. However, on the contrary, their final reports were written from a holistic view; that was, not only based on their own chosen academic fields but also from diversified viewpoints. Their final reports became very rich and impressive. This result indicates that our strategy was effective for cultivating a holistic point of view, which is thought to be essential for environmental leaders.

7.4 Self-evaluation of the Cambodia Unit

The Cambodia Unit was successfully conducted twice in 2011 and 2012, respectively. The Unit was found to be satisfactory for students according to our self-evaluation. The following sections discuss the evaluation of the Unit.

7.4.1 The Improvement on Lecture Preparation

The theme of this Unit was too broad for them to approach from a holistic viewpoint. As a consequence, in 2011, students found it difficult to initiate group work with a concrete idea under severe time pressure. Therefore, in 2012, we offered the preliminary lectures using a teleconferencing system. This enabled students to have clear ideas about the environmental issues in Cambodia in advance of the actual field visit.

In addition, through the collaboration with many organizations, the lectures were carefully chosen from different stakeholders; government sector, foreign government-related organization, non-governmental organization, profit and non-profit organization. The topics included various environmental issues, such as solid waste management, urban flooding, water management and urban development. As is mentioned earlier, one of the characteristics of the development in Cambodia is that it does not depend on a solo stakeholder. The lecture which cover various aspects relating to environmental problems were essential to let the students think from holistic viewpoint.

On top of the scientific knowledge, the lecture from a real environmental leader based on his experience was added in 2012, which enriched the program contents. This lecture highly motivated the students and let them think about the qualities of an environmental leader, including the sense of responsibility, braveness, transparency, vision and passion. Lots of positive feedbacks were given by students on this lecture.

7.4.2 Reschedule of Site Visit and Lecture

One of the differences between the Cambodia Unit 2011 and 2012 was that the program components within 1 day. In 2011, the contents of a lecture and a site visit in the same day were not necessarily related to one specific issue while those were in 2012. In other words, in each day in 2012, the students gained the knowledge from classroom lecture and site visit on a specific issue and had a time for group discussion on that issue in the same day. In this case, ideas and impressions of students were well exchanged and students were able to integrate their new knowledge with their background knowledge step by step.

In Cambodia Unit, two areas, remains of Pol Pot Regime in Phnom Penh and Angkor Wat in Siem Reap, were chosen as mentioned before in order to give students general background knowledge to have holistic understanding about environmental issues in Cambodia. In 2011, we started our fieldwork in Phnom Penh and moved to Siem Reap where we concluded the program. However, we found that the students preferred to do more research in groups in Phnom Penh after they studied the issues in Siem Reap, from the students' feedback. The site visit was rescheduled in 2012 accordingly. We returned to Phnom Penh after the visit to Siem Reap. It was one of the alternatives to let the students conduct problem analysis and create their own research structure.

7.4.3 The Future Challenge of the Cambodia Unit

One of the challenges in organizing the environmental leadership education program that we found in Cambodia was problem setting: how much the broad range of environmental issues have to be narrowed down. The lesson learned from the first Cambodia Unit was that the students tended to lose their way to find issues to be focused on since they were given too much information about broader environmental issues without any guidance. After the self-assessment of program, students in 2012 were given an assignment that has five specific topics to study and discuss in groups. However, this resulted in students selecting their group discussion topics only from the topics listed in the assignment. Furthermore, the students did not go far enough to explain how those topics are related at a fundamental level and answer to the core problem, although this method led students to analyze their chosen topic at a deeper level.

Another challenge is the impact that a leadership education program can bring over to a society. Education is an investment that needs time to get the fruits. Thus far, APIEL is successful in getting the various local stakeholders involved in this program. However, it will take another while for the participating students to be able to make significant contributions to solving the problems facing the Cambodian society.

7.4.4 Leadership Development in Students: Strive for Holistic View

In case of the program of 2012, good amount of basic information about Cambodia through preliminary assignment, and the lectures and visits enriched the knowledge that students already had and gave them stimulation to investigate further. This combination enabled students to address diversified aspects of environmental problems that teaching staff expected them to find from the study area.

The students in 2012, divided into four groups from (a) to (d), initiated the group discussion and eventually set their presentation topics by themselves. The topics for final presentations were as follows: (a) Capacity and Management of Irrigation in Cambodia, (b) Water Supply in Rural Area, (c) Role of ODA in Sustainable Livelihoods in Rural Cambodia and (d) Solid Waste Management in Phnom Penh. They were able to present concrete and feasible propositions for the sustainable development of Cambodia. The authors believe that they became aware how important it is to think out when they face difficult challenges. The authors also believe that a proposition has to be made not only from subjective, or single point of view, but also from objective and holistic viewpoints.

Acknowledgments The Cambodia Unit was successfully conducted with tremendous support from Royal University of Phnom Penh and Seoul National University. Special thanks are extended to Mr. Sour Sethy, Dr. Neth Baromey, Dr. Hyejin Jung and Dr. Kyoungmin Kim for their valuable input and comments.

References

1. International Monetary Fund (2012) World Economic Outlook. Washington DC, p. 194
2. World Commission on Environment and Development. "Our Common Future, Chapter 2: Towards Sustainable Development" http://www.un-documents.net/ocf-02.htm. Retrieved 28 September 2011. Accessed 6 March 2013
3. Princeton University Oral History Program (2009) Innovations for successful societies: Ek Sonn Chan. http://www.princeton.edu/successfulsocieties/content/superfocusareas/traps/RT/oralhistories/view.xml?id=157. Accessed 31 Oct 2012
4. Biswas AK, Tortajada C (2010) Water supply of Phnom Penh: an example of good governance. Water Resour Dev 26:157–172

Chapter 8
Resonance in the Asian Program for Incubation of Environmental Leaders

Neth Baromey, John Stuart Blakeney, Bayarlkham Byambaa, Ki-Ho Kim, Mingguo Ma, Jatuwat Sangsanont, Sour Sethy, Chettiyappan Visvanathan, Xin Li, and Yuan Qi*

Abstract The following accounts are from teaching staff at universities that have collaborated with The University of Tokyo in the APIEL's field exercises, as well as three alumni. The purpose of this chapter is to reflect the mutual influences of resonances

*All the authors contributed equally to the article and are listed alphabetically.

N. Baromey
Faculty of Social Science and Humanities, Royal University of Phnom Penh,
Phnom Penh, Cambodia

J.S. Blakeney
Ernst and Young, Harcourt Centre, Dublin, Ireland

B. Byambaa
Two-Step-Loan Project, Japan International Cooperation Agency, Ulaanbaatar, Mongolia

K-H. Kim
Asian Institute for Energy, Environment and Sustainability,
Seoul National University, Seoul, Republic of Korea

M. Ma • X. Li • Y. Qi
Cold and Arid Regions Environment and Engineering Research Institute,
Chinese Academy of Sciences, Lanzhou, China

J. Sangsanont (✉)
Asian Program for Incubation of Environmental Leaders (APIEL),
Department of Urban Engineering, Graduate School of Engineering,
The University of Tokyo, 7-3-1 Hongo, Bunkyo-ku, Tokyo 113-8654, Japan
e-mail: jatuwat@env.t.u-tokyo.ac.jp

S. Sethy
Faculty of Science, Royal University of Phnom Penh, Phnom Penh, Cambodia

C. Visvanathan
School of Environment, Resources and Development, Asian Institute of Technology,
Pathumthani, Thailand

T. Mino and K. Hanaki (eds.), *Environmental Leadership Capacity
Building in Higher Education*, DOI 10.1007/978-4-431-54340-4_8,
© The Author(s) 2013

created by Asian Program for Incubation of Environmental Leaders (APIEL) and collaborators across region, stakeholders and disciplines in Asia. The teaching staffs from Asian Institute of Technology, Cold and Arid Regions Environment and Engineering Research Institute, Chinese Academy of Sciences, Royal University of Phnom Penh and Seoul National University have shared their experiences with APIEL to foster future environmental leaders. Three APIEL alumni comments impact of APIEL on academic and professional development, its relevance to subsequent activities, and their own development.

Keywords Alumni • Career • Collaboration • Resonance

8.1 Collaboration with APIEL in Thailand Unit: As We Look Ahead int. the Next Century[1,2]

Leaders Will Be Those Who Empower Others—Bill Gates

8.1.1 Introduction

As Asia is composed of many highly populated countries at differing stages of development, the key is to build a sustainable society within the inherent constraints of the earth. An interdisciplinary, multifaceted approach is essential to tackle global environmental issues, which are becoming more complex, varied and are deepening. Initiatives confined to regions cannot resolve global-scale environmental problems. At the same time, however, local perspectives are crucial for resolving global issues, and regional idiosyncrasies, social conditions, political systems, and cultural backgrounds must all be taken into account.

The APIEL was designed to build broad and resonant networks among universities, research institutions and program graduates throughout Asia. For the APIEL Thailand Unit, two leading universities in Thailand—the Asian Institute of Technology (AIT) and Kasetsart University (KU)—worked jointly with The University of Tokyo (UT). The unit's objective is to produce individuals who attain knowledge, skills, and competencies and who can play a major role as environmental leaders in various settings, such as educational research institutions in Asia or in other parts of the world and in global organizations and corporations.

---

[1]This sub chapter is written by one of the authors, Chettiyappan Visvanathan, from School of Environment, Resources and Development, Asian Institute of Technology, Thailand.
[2]Further information about Thailand Unit can be found in Chap. 4.

8.1.2 Partnering Institutions

AIT promotes technological change and sustainable development in the Asian-Pacific region through higher education, research and outreach. AIT was established near Bangkok in 1959, and has become a leading regional postgraduate institution and is actively working with public and private sector partners throughout the region and with some of the top universities in the world. Specifically, Environmental Engineering and Management and Water Engineering and Management partnered in the course with APIEL over 4 years for knowledge and experience sharing from AIT.

KU is a public university where bodies of knowledge and research potential have been continually accumulated for nearly seven decades. Today, it is a national research university endorsed by the Commission on Higher Education of Thailand with the vision to become "the world's leading research university in agriculture, food, technology and innovation." The Faculty of Engineering in KU is one of the top ranking institutes of engineering in Thailand. It offers bachelor, master, and doctoral programs in 15 fields including Department of Environmental Engineering. It educates engineers to serve society, develop the country, and pursue excellence in engineering and scientific advancement.

8.1.3 Thailand Unit Conducted Over the Years

UT, AIT and KU have collaborated to tackle Asian environmental problems in a tropical environment. This unit ensures student exchange between the three participating institutes covering an interdisciplinary program with from 10 to 20 students from more than ten countries.

From 2009 to 2012, major sectors of environmental problems have been addressed in the joint program. In 2009, issues on sustainable solid waste management in Asian developing countries were taken up. By 2011, another important issue on sustainable urban water use and management was addressed. Flooding is a natural phenomenon and various human activities as well as climate change have aggravated the problem, causing economic loss. The nationwide flooding in late 2011 resulted in huge losses in Bangkok and other urban areas. Thus in 2012, Thailand Unit focused on the issue of urban waste management with a special focus on flood management in Bangkok.

The series of Thailand Unit demonstrated that students from different nationalities and academic backgrounds could sit together and review these issues from various angles, within the technical, economic and social context. The unit revealed the importance of incubating new environmental leaders with negotiations skills for considering the importance of "economic growth along with environmental protection." This can be achieved only by providing these leaders with communication and management skills. In this way, complex issues can be expressed in layman's language to the public, and support can be gained for environmental protection.

This specially designed, short but focused, unit provided the best format for the students to understand the problems by equally mixing classroom lectures, field-work and sampling, as well as community consultations. This, nontraditional format educational concept is one of the aspects most highly appreciated by all the students who took part.

8.2 Collaboration with APIEL in Oasis Unit[3,4]

Cold and Arid Regions Environment and Engineering Research Institute (CAREERI), Chinese Academy of Sciences, is the Chinese collaborative partner in the APIEL for Oasis Unit (see Table 8.1). CAREERI is a large research institute that has seven laboratories and three research centers, highlighting the characteristics of the glacier, desert, and alpine areas of ecological environments and the sustainable development of resources.

The memorandum of understanding on academic cooperation between the Graduate Program in Sustainability Science, UT, and the Laboratory of Remote Sensing and Geospatial Science, CAREERI, was signed on 31 December 2008. The APIEL Oasis Unit was designed and developed at that time. Since then, for 4 consecutive years, joint fieldwork was performed in the Heihe River basin, the second largest inland river basin in China. The river basin is also an important scientific research base for CAREERI and even for the whole country. Forty-one students from 14 countries in total have joined the joint fieldwork during the past 4 years. There are 14 students and 4 professors from CAREERI that have joined the fieldwork (Table 8.1). Students and teachers from CAREERI are familiar with the Heihe River basin; they supported the members from UT by providing relevant data, information, and suggestions.

Another important partner of the joint fieldwork comes from the Zhangye Water Authority. In particular, Director Tuo Xingfu, Vice Director Liu Guoqiang, Luan Limin, Zhang Wenwu, and Liu Xiaojun from the water authority contributed to the fieldwork significantly through their active participation and full support. The local water resource managers have a deep understanding of local environmental problems, and they also have abundant experiences dealing with those problems.

As Chinese-side participants, CAREERI students take the role of collaborators during fieldwork. Because there are some differences between their studying specialties and the investigative contents of the fieldwork, the improvement in specialized knowledge is relatively limited. But our students have benefited a great deal from the following two perspectives:

1. *English conversation and application ability*: This is a distinct opportunity for CAREERI students to use their English in real-world situations. UT students

---

[3]This subchapter was written by three authors of Mingguo Ma, Xin Li and Yuan Qi from Cold and Arid Regions Environment and Engineering Research Institute, Chinese Academy of Sciences, China.

[4]Further information about Oasis Unit can be found in Chap. 5.

Table 8.1 Students and professors from CAREERI who have participated in the Oasis Unit

Year	Students	Teachers
2009	Song Yi, Tan Junlei, Wang Haibo, Yu Wenping	Prof. Li Xin, Prof. Ma Mingguo
2010	Song Yi, Tan Junlei, Wang Xufeng, Hu Wenbing, Yu Wenping, Liu Chao, Fang Miao, Zheng Zhong	Prof. Li Xin, Prof. Ma Mingguo, Assoc. Prof. Qi Yuan, Assoc. Prof. Ge Yingchun
2011	Wang Xufeng, Wang Haibo, Xu Fengying, Xie Yanmei	Prof. Li Xin, Prof. Ma Mingguo
2012	Wang Xufeng, Jia Shuzheng, Xiao Lin, Wang Hongshu	Prof. Li Xin, Prof. Ma Mingguo

come from all corners of the world and English is the common language for communication. There are few opportunities for the Chinese students to use English so often and intensively.

2. *Exchange of learning styles*: The learning style of APIEL is itself distinctive and worth learning. For example, the professors give the fullest play to students' initiatives. Discussion is one of most important learning tools in the fieldwork. The Chinese students can use these learning and communication skills developed during fieldwork in their future study and work. The effect is especially clear when our students join the fieldwork and training classes with UT students.

From my point of view, the Oasis Unit has the following three features. They are valuable experience and lessons. They are important for us to develop a new cooperation plan.

1. *Before hand preparation and field survey*: It is better to see once than to hear a hundred times. Even though the students obtained information about the research area by reading papers and presentations before they visit the Heihe River basin, this information is still abstract. It is very difficult for UT students to prepare research topic and survey plan reflecting real-world situation. In the first 2 days of the fieldwork, students are guided to visit some places along the Heihe River, as well as the local water authority and the irrigation channel system. In this way, they can gain more information about this region, such as government policy, local customs, and potential difficulties with their upcoming work. After the 2-day visit, students usually find some problems with their original plan, and groups are divided according to their interests. Professors give their advice when students make their plans.

2. *Data collection through experiments and interview survey*: Usually, students have 4 days to collect data and information to support their research. In this process, they need to communicate with local people, such as farmers, government officials, scientists, and so on. Students from CAREERI play an important role in this process because of language limitations.

3. *Getting to a conclusion and giving suggestions to the local water authority*: Based on the collected data and information, students are asked to arrive at a

conclusion and give suggestions to the local water authority about how to solve the environment problem that they find. The local water authority also gives feedback and comments to the students. Although field experiments make a limited contribution to solving local environmental problems, the students' ability to find and solve problems has improved.

The fieldwork is in the process of improving. The fieldwork had been performed during the summer from 2009 to 2012 in the Heihe River basin. There are some big differences between these four fieldwork exercises. Knowing these differences is important for us to develop a new cooperation plan. On one hand, beforehand preparation is very important for a successful fieldwork. Especially, it is important to find concrete research topics, which can save time and greatly improve working efficiency during field survey. The students who joined fieldwork in 2009 were better prepared than the students of other years. On the other hand, the active participation of CAREERI students is important for a successful fieldwork. The Chinese students just acted as guides in 2009. However, they participated in the fieldwork fully in 2010–2012. They discussed and worked together with UT students during the whole process from preparation, survey, until final report writing.

Based on the comparisons on these four times of field exercise units, the following suggestions can be made to improve collaboration with UT.

1. The preparation of the field exercise unit can be improved by making a feasible proposal that can be applied on site by both UT and CAREERI students. The proposal needs to include the background, purpose, research content, experimental design, and anticipated results. Both UT and CAREERI students need to spend enough time to prepare this proposal during the preparation phase.
2. There were three different research topics in each year and two study areas in 2010. One common research topic is enough for each fieldwork although the investigation or measurement can be divided into two or three parts. However, different parts should be around the same topic. The students should try to obtain more samples of the questionnaire survey and the measurements, which can improve the persuasiveness of conclusive results and suggestions to the local water authority.
3. A Chinese student from UT side is necessary for each student group. A Chinese student can offer a lot of help to the other students during the whole process of the fieldwork program. For example, background information can be obtained more easily because of the limitations caused by language barrier.
4. Comparisons of the methodologies and results can be carried out among the different fieldwork areas.
5. If the survey contents are close to the specialties of the Chinese students, our students can be more involved in the preparation work. If possible, these students could participate in making the fieldwork plans in Japan during the preparation stage.

8.3   Collaboration with APIEL in Cambodia Unit[5,6]

8.3.1   Introduction

The Royal University of Phnom Penh (RUPP) is the oldest and the largest public university in Cambodia. RUPP first opened its doors as the Royal Khmer University on 13 January 1960, with the National Institute of Judicial and Economic Studies, the Royal School of Medicine, a National School of Commerce, the National Pedagogical Institute, the Faculty of Letters and Human Sciences, and the Faculty of Science and Technology.[7] The language of instruction during this period was French.

However, today it is unique in Cambodia for offering specialist degrees not only in fields including the sciences, humanities and social sciences, but professional degrees as well in fields such as information technology, electronics, psychology, social work, tourism, and the environment. It also provides Cambodia's foremost degree-level language programs through the Institute of Foreign Languages. RUPP has many achievements, and it now has full membership in the ASEAN University Network. The APIEL Cambodia Unit was initially set up in 2011.

8.3.2   Institutional Involvement in the Cambodia Unit

The Cambodia Unit is organized by three universities: UT, Seoul National University (SNU) and RUPP. It has organized two field trips/lecture programs that were held in September 2011 and in August 2012. The unit was designed based on experienced, professional skill development, and lessons that the universities use to effectively and efficiently instruct the students. Various activities including lectures from experienced resource persons, fieldwork, group work and group presentation are used.

8.3.3   Achievements of the Cambodia Unit

The program was designed to follow a participatory, problem-solving approach. There is a combination of activities such as lecturing and fieldwork. Through the

---

[5]This subchapter is written by two authors, Sour Sethy from Faculty of Science and Neth Baromey from Faculty of Social Science and Humanities, Royal University of Phnom Penh, Cambodia.

[6]Further information about Cambodia Unit can be found in Chap. 7.

[7]Further information about RUPP is found in their website (http://www.rupp.edu.kh/content.php?page=about_rupp).

involvement of the various activities and evaluations by Cambodian students, this program brought significant benefit to participants, especially those from RUPP.

8.3.3.1 Development of the Training Program

The training program was developed by UT in discussion with SNU and RUPP, and through a series of discussions and meetings it was decided to base it on a participatory approach with professional (leadership) training. Therefore, the program has been implemented effectively and successfully. Participants in the Cambodia Unit were able to learn many things, not only from a technical perspective, but also about the social and multicultural components. This design provided us with ideas on how to successfully coordinate with stakeholders from different nationalities. It was also found that the participants, particularly the Cambodian students, absorbed good lessons, learning how to manage their time, and improve their learning style, research skills and English language ability.

8.3.3.2 Fieldwork

From the hands-on practices, RUPP students gained deeper knowledge on sustainable development in Cambodia. They basically preferred to have practical training and visits to project sites. They also said that the practice at the sites was complementary to the theories and that the concepts were easy to understand, thus they were interested and they found the program enjoyable. They were able to get more understanding of the real situation and lifestyles in local communities. Furthermore, the program provided a valuable opportunity for Cambodian students, who usually have limited chances for practical training. The students enjoyed this approach and were willing to become involved in the fieldwork. They wished that the unit could be available to other students, as well.

8.3.3.3 Cooperation in the Future

The Cambodia Unit has been active only 10 days per program including the lectures and fieldwork, but the students have felt that there were positively results. The Cambodia Unit has also contributed to a positive outcome for our society, as well as sustainable development, through the problem-solving approach, especially on the environment, society, and the economy. Importantly, the Cambodia Unit will be an initial point and have continuing good opportunities for further cooperation of the relevant stakeholders in the future.

   RUPP wishes the unit will continue its mission for the sake of sustainability and for producing more young environmental leaders for this region as well as for the world. RUPP is ready to continue to cooperate with the unit as well as with UT and SNU. The perspective of RUPP is that the Cambodia Unit should continue and should engage more stakeholders from different countries. In this way, the unit's vision will be widely shared and make both a global and region contribution. Additionally cooperation in the future should be not focused only on a short training program, but should also consider long-term joint research and capacity building for continuing education.

8.3.4 Conclusion

The unit was based on a problem-solving approach, which was popular with the students and the course designers, thus the students found it more attractive to be involved. Based on this approach, the students learned many things, such as the technical, social, and cultural aspects. The key achievements of the unit included:

1. The students who participated absorbed various information, particularly concept of sustainability related to social, environmental and economic factors. These give them ideas on how to help their countries develop sustainably.
2. Networking for the future was started with the unit. RUPP's APIEL alumni association was also established by the students themselves. Additionally, the initiatives of the Cambodia Unit will be a starting point for networking and cooperation in the future.

8.4 Two Years Collaboration With APIEL[8]

8.4.1 Introduction

For the past 2 years, the Asian Institute for Energy, Environment and Sustainability (AIEES) at SNU has been given a great opportunity to participate in APIEL and send students as well as professors to lead the students during the field trips for APIEL. One of the reasons why we have participated in this program is that AIEES is also planning to build an Asian green leadership center at AIEES in the years to come. In this regard, since UT started APIEL earlier, by participating together, we wanted to learn how UT is operating their international education program.

Two years ago, SNU hosted the Asian University Conference for Green Leadership. For this conference, we invited Professor Hanaki, where he introduced the APIEL to us. And we have been participating ever since. Professor Hanaki also took part in a seminar prior to the establishment of AIEES back in 2008. Therefore, AIEES has had a deep connection with him, and based on this relationship of trust, we have collaborated with APIEL.

---

[8]This subchapter is written by one of the authors, Ki-Ho Kim, from Asian Institute for Energy, Environment and Sustainability, Seoul National University, Republic of Korea.

8.4.2 Current Situation

AIEES at SNU has participated in APIEL three times, starting from the Cambodia Unit in August 2011. In the Cambodia Unit, the local partner universities were RUPP, and in the Greater Pearl River Delta (GPRD) Unit, Hong Kong University of Science and Technology and Sun Yat-sen University. Each of these universities has provided graduate students and lecturers.

From SNU, 20 students have taken part so far. Students with diverse majors, such as international studies, environmental studies, natural science, engineering, law, and agricultural science, have participated in the field program for cultivating leadership concerning the environment and returned good feedback.

8.4.3 Major Elements of Education

In the Cambodia Unit, the program mainly covers topics, from infrastructure to environmental management issues—waste disposal, recycling, and water quality protection—and SNU has focused on rural development and capacity building as well as contributing to physical and social capacity development for sustainable development.

The GPRD Unit provided an opportunity to study several components and techniques for a more sustainable urban plan and urban design. In particular, the students' practical exercises on mapping and gaining techniques for handling vast amounts of information related to cities greatly helped them come to arrive at the future direction of sustainable urban development.

In each unit, site visits to understand the local culture and history, as well as interviews with local residents, have provided a good opportunity for the students to draw a profound understanding of other cultures. The cooperation that the students from each university showed helped promote the understanding and capacity to produce definite outcomes on site. It is also significant that opportunities to directly experience the possible direction of sustainable development, rather than in theory, were provided to graduate students from each university through this program.

Each unit covered fields related to sustainable development, such as waste, sewage disposal, air quality management, urban planning, eco-tourism, and official development assistance from the Korean International Cooperation Agency and the Japan International Cooperation Agency. Through lectures from experts and field-work, each program was able to cover general plans and implementation measures for sustainable development in developing countries.

Furthermore, in each group, made up of graduate students from each university, exchanges were held for carrying out the research project: establishing strategies for sustainable development. These exchanges in combination with in-depth research under the instruction of the resource person for each group progressed to outcomes in forms of presentations and recommendations.

8.4.4 Thoughts on the APIEL

8.4.4.1 Strengths

"Concerning the management of APIEL, what has impressed me the most is that you have a full-time professor and administrator for this program. I think they are professionally and academically excellent. Whatever we are doing at universities, the most important thing seems to be the people who are operating the program. In the case of APIEL, the leader is Professor Keisuke Hanaki, who has a global perspective and great vision for the program. Also, the performance of supporting staff including faculty members and administrators has been excellent. If this program did not have those people, I think it would not have been managed in such a thorough professional manner."

"For the content, as a participating professor who looked at and examined the participating students in the Cambodia and GPRD Unit, APIEL provided the students with good opportunities to promote 'hands-on knowledge' related to sustainability issues through the experience of participating in the unit. Through the experiences, they are expected to grow into more mature leaders."

"In particular, the students worked in groups, enhanced communication, and engaged in vigorous discussion until late in the night with other students from different backgrounds and cultures to produce definite outcomes. This allowed the students to understand others who have different opinions and learn the attitude required of a leader to produce an outcome. Thus, the Cambodia Unit offered a very good platform for students to foster leadership. Another strength is that the unit was a melting pot for students of diverse nationalities, majors, and races, not only from SNU but also UT and RUPP."

"It is doubtful whether it is possible to find facts and suggest solutions in a period of only 2 weeks or how much meaning there could be in such a solution. However, this program is very significant in that it allows the students to genuinely ponder the problems and become equipped with the technical capacity and passion to solve them."

8.4.4.2 Weaknesses

"Regarding the education program provided, there was a lack of information concerning the site and there was realistically not enough time and process to sufficiently understand what was available. As a result, we had insufficient responsiveness to the field activities. Moreover, although the scope of the program for such a comprehensive subject as sustainable development seemed to be adequate on the surface, it turned out that there was limited time to engage in in-depth discussions."

"Also, there was a disharmony between the lecture provided and the resolution of problems in the subject. The fieldwork had limitations, and lectures are composed and arranged according to the major subjects of the participating professors rather than the actual subjects."

"Regarding the development of leadership, although SNU students might have a high level of academic understanding, there were times when I was skeptical about whether they had the capability for leadership in this era of climate change and sustainability."

8.4.4.3   Suggestions

"At SNU, there is a similar program called the Green Leadership Program (GLP) at the undergraduate level to increase the students' environmental awareness and leadership skills. GLP is provided for 5,000 undergraduate students at the university, regardless of their major. Sixteen new courses have been created for the program. At the end of 4 years, when a student takes 15 credits, or five courses, which corresponds to about one-tenth of the total 140 credits required for graduation, the student will receive a green leadership certificate along with a bachelor's degree. After graduation, we expect the student can be not only a philosopher, economist, sociologist, and architect, etc., but also a green leader. It has been 3 years now since we began this program. By the end of the 2012 academic year, it is expected that we will begin to produce the first students to receive the green leadership certificate."

"Although of a different character, I believe GLP at SNU and APIEL at UT can learn from each other. Of relatively recent birth, I think GLP can learn a lot from the maturity of APIEL. On the other hand, though a great program already, APIEL may want to seek to improve the present program towards a greater diversity in terms of the students and subjects the students are taking. Although the two programs may seem to be different, I think they provide possible future directions on how each program could evolve into. Therefore, the character of each program could be complementary, and as a result, I think the two can learn from each other."

8.5   Comments from Alumni

8.5.1   Impact of APIEL on Academic and Professional Development[9]

> Nothing else but an apple grows on an apple tree. Likewise, everything in this world is interrelated and one becomes a root cause of another.

Metaphorically, the author considers APIEL as an "apple tree" of her current progress, as it has grown an "apple seed" inside her. This is because the author has

---

[9]This subchapter was written by one of authors, Bayarlkham Byambaa, graduated from Graduate Program in Sustainability Science, Graduate School of Frontier Sciences, The University of Tokyo in Year 2011.

had a number of opportunities to join and become a member an APIEL unit, which has provided her with valuable experiences and practical knowledge.

First, the author had a chance to join the Oasis Unit 2010, which was held in the middle and lower reaches of the Heihe River basin in China. The field trip was fruitful, as the author learned many things and practiced leadership skills throughout the whole unit. The field exercise was very challenging and stimulating in the sense that it gave her a unique chance to carry out and pursue her own interests with real practice. It has changed and strengthened her views and knowledge towards the environment and human relationships, and has motivated her a great deal to contribute to environmental conservation. The lessons and skills the author has acquired are very much essential to her current work as an environmental specialist and as an activist in her home country, where there are many similar stories to the story of watershed management in the Heihe River basin.

The experiences the author has gained through fieldwork has broadened her understanding and knowledge of water environments and made her confident to apply the same approach and methodologies to her own research. A specific example of a knowledge application from the Oasis Unit was a water quality measurement analysis and assessment for her master's thesis field research. Before joining the unit, the author worried about her weak knowledge and understanding of water environments as the author is from a social sciences background. However, the author was always interested in becoming familiar with water quality issues and the author was longing to conduct research using an interdisciplinary methodology, especially for water quality sampling in the field. Luckily, with the support and valuable guidance of the professors and experts in the field throughout the unit, the author overcame her concerns not only through the fieldwork, but also it made it possible for her to realize interests in practice later on by applying water quality measurement analysis into her own field research in the Zaamar goldfield in Mongolia.

Most importantly, environmental leadership education exercises have motivated her to contribute to finding solutions for environmental problems by taking the initiative and doing her best in her future career. Experience from the Oasis Unit has directly influenced her thoughts and actions in helping to preserve the water environment in her home country. After returning to Mongolia, the author has reorganized and activated NGO activities and has launched a small project for creating a drinking water quality database at the provincial level in the South Gobi Province. From the project, environmental officers at 15 *soums* (districts) in South Gobi were provided with simple water quality testers and some knowledge and training for drinking water quality testing. Local people are now able to conduct monthly measurements at major sources of drinking water and submit their results to the monitoring database at the provincial center. The project is a pilot program; it has being implemented in only one province, however, it is a pioneer in the country where there was no permanent and long-term database for drinking water quality.

Second, as an APIEL alumna, the author was fortunate to join the Coca-Cola Young Environmental Leaders Summit in Hokkaido on March 2012. The lessons learned and skills acquired from the week-long intense workshop on corporate social responsibility and the practicum on the project design matrix are very much essential to her current work in Mongolia for the Japan International Cooperation

Agency's Two-Step Loan Project for Small and Medium-Scaled Enterprises (SMEs) Development and Environmental Protection. The author mostly works with SMEs to promote long-term business loans and to help them develop and design good and bankable projects. As well, the author assesses the environmental impact and socio-economic benefits of the projects.

Another valuable opportunity provided to her because of her APIEL experience was an internship at the Asian Development Bank Institute (ADBI) in Tokyo. During the 2-month internship at the ADBI, she joined the Climate Change and Green Asia Flagship project team and has completed a country report on climate change mitigation for Mongolia. Thanks to this internship and helpful guidance from her supervisor, the author was able to learn about climate change issues in Mongolia, a topic that she was not so familiar with. Through reviewing and analyzing official documents and literature on climate change mitigation in Mongolia, as well as the development trajectory and emissions, the author has assessed the correlation between CO_2 emissions and economic growth by analyzing the factors shaping the current emission scenario profiles, projected emissions, and economic growth in 2020. The author has developed recommendations for further policy action. Overall, her experience has deepened her understanding of climate change issues with a special focus on climate change mitigation in Mongolia.

A recent opportunity because of her APIEL experience, which was grown from the internship opportunity at ADBI, was participation in one of the ADBI jointly organized events, a sub-regional workshop on Millennium Development Goals and Post-2015 Development Agenda in the Central and East Asian countries in Almaty, Kazakhstan, in September 2012. The author had a chance to join the discussion on regional socio-economic and environmental development issues.

APIEL has given her a lot of opportunity, and it still does in fact. After such beneficial experience and all the lessons the author has received through APIEL programs, it became obvious that her future work and career would be related to environmental leadership. The author feels that she owes a lot to APIEL for contributing to the sustainable development of Mongolia and the region. All in all, APIEL has helped to form her ambition and vision to work in the environmental sustainability field, while keeping a good balance between academic and practical views. As time goes by, "apples" will be growing even tastier.

8.5.2 APIEL Experience and Its Relevance to One Student's Subsequent Activities[10]

In this account, an alumnus discusses how his experiences in APIEL fed into his current career and future career plans. The author is currently training as an

---

[10]This subchapter was written by one of authors, John Stuart Blakeney, graduated from Department of Urban Engineering, Graduate School of Engineering, The University of Tokyo in Year 2011.

accountant with a large professional services firm and volunteering with the Irish National Trust. The APIEL includes students from diverse backgrounds with the author's undergraduate degree being in geography, a subject that explores the tension between human activities and development, and the natural environment, and thus led naturally to an interest in APIEL.

8.5.2.1 The APIEL Experience

"There was a core taught element to the APIEL which familiarized students with some of the main environmental issues which are being grappled with (particularly in Asia) and provided analytical frameworks with which such issues can be studied. Whilst it was important to have an understanding of the science behind the issues, with the main focus on environmental leadership, a key learning objective was to understand how stakeholders could be mobilized and managed to achieve particular objectives.

"However, it was the fieldwork element of APIEL that had the greatest impact on the students' development. The author took part in one overseas fieldtrip in the Pearl River Delta in China where students from a great diversity of backgrounds had the opportunity to network. The theme of the fieldwork was trans-boundary pollution. Given the current standoff on environmental issues between the developed and developing worlds, it was particularly fascinating to get first-hand insights into how political issues are dealt with in the Chinese context. This included privileged access to sites of pollution, and policymakers and business leaders who provided details of their own agenda and their vision for how environmental issues will be dealt with in the coming years and decades.

"Other students took place in field units in other parts of Asia, such as Thailand, Vietnam and Northern China. The following year participants from each of these field units were brought together at the Young Environmental Leaders Summit held in Hokkaido. This workshop focused on corporate social responsibility (CSR) and particularly on how, in recent times, companies have sought to integrate CSR into their core business strategies, rather than using it in a responsive way to manipulate public opinion.

"The Green Energy Workshop included the opportunity to examine the energy challenge faced by Japan. As part of this trip students visited a nuclear power plant and a nearby pump-storage plant and learned how the two are used in tandem so that the excess power generated at night by the nuclear plant can be used to provide for generation at peak daytime hours. Debates held during this unit provided students with a unique perspective on the controversy surrounding the future of the nuclear industry given the context of the Fukushima nuclear disaster, which shortly followed the field unit.

"By examining such tough issues in multidisciplinary groups, those from non-engineering backgrounds were forced to accept that ideology is constrained by what technology will allow. At the same time, those from purely technical backgrounds gained an understanding of externalities and the conflict that can exist between achieving efficiency and economies of scale in production processes, whilst maintaining democratic accountability and broad access to the benefits of technology."

8.5.2.2 APIEL and Future Career Opportunities

"Although the program has been running for just a few years, the APIEL boasts alumni in a wide range of fledgling careers. The author is determined to remain engaged with environmental issues and particularly with developments in how we power our society. Having had the opportunity to interact with businesspeople and politicians through APIEL, the weight of money and economics in the discussion of all the issues with which students were faced was apparent. As such, in order to have a real impact, the author felt it was vital to be able to understand the imperatives of the business community and this has led him to take a position as a trainee accountant. This is something he hopes to leverage in the future in either a leading business or regulatory role in the energy sector.

"Roles in business are available to those from non-business academic backgrounds provided they demonstrate the necessary qualities. Given the economic climate, the job market in Ireland is extremely competitive but recruiters have proven highly receptive to international experience and particularly to that gained during on APIEL. Working on projects in interdisciplinary, multicultural teams is a great challenge, and this is something that is acknowledged by businesses. Furthermore, by equipping young people with such skills and sending them to the global workforce, the program can have strong hopes for an extremely powerful alumni network which will reinforce its importance.

"Active engagement in civil society is something that should be strongly encouraged through APIEL. The author has joined the Irish National Trust which he hopes will allow him to stay in touch with and influence environmental policy even in the immediate future. Again having alumni involved in NGOs and charity organizations will provide APIEL with a great pool of experience to draw on."

8.5.2.3 Recommendation

"There is perhaps one central facet in which the program might be further improved. Encouragement, facilitation and perhaps compulsion of students to take on internships and work placements would be invaluable in allowing them to link into the workforce and community on completion of their studies. Organizers should endeavor to develop strong relationships with a diverse range of institutions who could accept students to such positions. Such placements would also amplify the benefits of the Young Environmental Leaders Summit as students would be capable of bringing still more to their project groups.

"It is a credit to the largely young group of academics who have powered APIEL behind the scenes that it has come so far in such a short space of time. Their enthusiasm and effort has meant that the foresight of the Japanese Government in investing in such a program has already paid rich dividends. Provided the same energy is dedicated to incremental improvement of the content and maintenance of funding for APIEL, it promises to be a forum that mediates and shapes opinion on environmental issues in Asia and further afield."

8.5.3 The Experience of APIEL and its Impact on His Development[11]

The APIEL was first organized when the author was studying at the Department of Urban Engineering, UT. At that time, knowledge limited to one specific academic environmental field was not enough; broad global environmental issues are becoming more and more important. APIEL could not be a more suitable and beneficial program for students in the environmental field. The APIEL curriculum not only consists of compulsory courses, which lets the students have a better understanding and analysis of environment-related problems in Asia but also consists of hands-on experience to propose environmental solutions in the region where actual environmental problems are occurring. Furthermore, the APIEL focuses on establishing, through collaborative projects, a network of people in environmental fields. This is one of the keys to success in resolving global environmental issues. The APIEL curriculum, which aims to incubate students to have the skills to solve environmental problems based on an interdisciplinary approach and specialized skills, really fascinated the author, making him decide to apply. At the end of the program, his decision could not have been more right as he has learned, accomplished, and gained a great deal.

The compulsory course for APIEL was especially meaningful. Students learned about the global aspects of environmental issues, focusing on Asian countries such as Japan, China, Korea and Hong Kong. The content was not limited to one aspect but came from various aspects. It helped them create a structured methodology for problem solving based on interdisciplinary knowledge gained from the APIEL curriculum and their own specialized skills from their academic fields. The integral knowledge was crucial for analyzing and solving global environmental issues and the course equipped them with it.

APIEL also provides students with a good experience in fieldwork, such as the Oasis Unit in 2009, which offered participants a chance to develop environmental leadership skills through education and from different disciplines. In the fieldwork for the Oasis Unit in Lanzhou Province in China, the focus was on sustainable watershed management in arid regions. During the program, the attendees developed their environmental leadership skills through three main parts: lectures, group work, and fieldwork.

Before the fieldwork, a lecture was delivered by a professor from CAREERI. The professor was highly knowledgeable about the environmental problems in the

---

[11]This subchapter was written by one of authors, Jatuwat Sangsanont, graduated from Department of Urban Engineering, Graduate School of Engineering, The University of Tokyo in Year 2011.

study area. Students gained an understanding and much insight into the environmental issues, and were then capable of constructing their own strategy to identify and solve problems. The group work was done before and during the fieldwork in China. They had a chance to work with a diverse range of students from different backgrounds and countries. They also learned how to exchange their knowledge and ideas with other people from different academic backgrounds. They broadened their perspectives through discussion. During the fieldwork, they had a field survey, visited with the government officers, and talked with local stakeholders. This was a valuable opportunity to get hands-on experience dealing with environmental problems. After the fieldwork, APIEL provided a platform for the students to present their proposals to the government officers, who have the authority to make policy changes, and to get comments from them. From these comments, the students could better understand the weak points or impractical parts of their proposal. This was really interesting and challenging for them as students. This experience helped them develop as environmentalists for the future.

Moreover, the students had a chance to participate in another unit: the Coca-Cola Young Environmental Leaders Summit. This unit was held for current participants and alumni. Participants from more than ten countries with different educational backgrounds came together to discuss the creative corporate social responsibility approaches. The Coca-Cola Unit challenged students to develop innovative solutions for a topic that they were not familiar with. This summit was especially beneficial for them to develop communication and negotiation skills. In addition, they were able to build up their own worldwide network with young environmentalists from different countries. This network will be valuable and beneficial for students in the future.

The experience from APIEL had a great influence on students attending the program, including the author for his career. He would like to be involved with environmental issues in Asia, especially in the Southeast Asia region, where his home country is located. After he got his doctoral degree in 2011, he started working as a project researcher under the APIEL in the university. His work is concerned with facilitating environmental educational programs. He needed to construct programs in Asian countries with professors to help educate and incubate students to become environmental leaders. His work focuses on such programs as sustainable water management in Thailand and sustainable development in Cambodia. The fruitful experiences from participating in APIEL really helped him to achieve in his job as a facilitator. For instance, having a wide environmental perspective and understanding the students' points of view, as well as communicating and negotiating were all essential for him to fulfill his task as an environmental facilitator and educator. In his career, he hopes to contribute to work on a global scale in the sectors of renewable energy, environmental health, and water pollution, etc.

In conclusion, the APIEL provides the participating students with a valuable multidisciplinary curriculum, practical experience, and a broad network related to environmental issues. Through the program, they are challenged to develop integrated solutions to environmental problems. It helps reveal their weak points, such as the knowledge and communication skills required for discussions and negotiations with people from different academic backgrounds. They are also motivated to promote sustainability of the environment and of society. Similar to other students, the author really enjoyed and appreciated the program. He felt thankful to have an opportunity to participate in this program and he hopes to work toward a sustainable future.

Index

T. Mino and K. Hanaki (eds.), *Environmental Leadership Capacity
Building in Higher Education*, DOI 10.1007/978-4-431-54340-4,
© The Author(s) 2013

Printed by Books on Demand, Germany